girl wide web

Intersections
in Communications
and Culture

Global Approaches and Transdisciplinary Perspectives

Cameron McCarthy and Angharad N. Valdivia
General Editors

Vol. 11

PETER LANG
New York • Washington, D.C./Baltimore • Bern
Frankfurt am Main • Berlin • Brussels • Vienna • Oxford

girl wide web

Girls,
the Internet,
and the Negotiation
of Identity

EDITED BY
Sharon R. Mazzarella

PETER LANG
New York • Washington, D.C./Baltimore • Bern
Frankfurt am Main • Berlin • Brussels • Vienna • Oxford

Library of Congress Cataloging-in-Publication Data

Girl wide web: girls, the Internet, and the negotiation of identity /
edited by Sharon R. Mazzarella.
p. cm. — (Intersections in communications and culture; v. 11)
Includes bibliographical references.
1. Teenage girls—Psychology. 2. Internet. 3. Adolescent psychology.
4. Identity (Psychology). I. Mazzarella, Sharon R. II. Series.
HQ798.G525   305.235'2—dc22   2004014680
ISBN 978-0-8204-7117-4
ISSN 1528-610X

Bibliographic information published by **Die Deutsche Bibliothek**.
Die Deutsche Bibliothek lists this publication in the "Deutsche
Nationalbibliografie"; detailed bibliographic data is available
on the Internet at http://dnb.ddb.de/.

Cover image by Ellie Brown

Cover design by Lisa Barfield

The paper in this book meets the guidelines for permanence and durability
of the Committee on Production Guidelines for Book Longevity
of the Council of Library Resources.

To my eight-year-old niece,
Amanda Rose Mazzarella, who,
through her example, inspires me
to embrace life and celebrate
the wonders of being a girl.

Contents

Acknowledgments

As I write this on September 8, 2004, it is hard to believe that it was barely over a year ago that I asked a fantastic group of scholars if they would be interested in contributing chapters to this volume. And here we are now with the book just about to enter the production process. So it is only fitting that I begin by acknowledging the contributors to this book. It was their enthusiasm for the project, intellectual expertise, and overall good humor that enabled this project to come together so quickly. It has been a genuine pleasure to work with each of them. I would like to extend a special thank you to one of the contributors, Susan F. Walsh, who organized what ended up being a very successful panel on girls and the Internet at the 2002 meeting of the National Communication Association. It was that panel that gave me the idea for this book. The timeliness of the topic, the intellectual engagement of the panelists and audience members, and the fun I had being a part of it all contributed to my conviction that exploring the relationship between girls and the Internet would make a great book topic.

Sharing that conviction since the beginning has been Damon Zucca, my editor at Peter Lang Publishing. From our very first contact about this project, Damon outwardly expressed both his excitement about the topic and his belief that the outcome would be successful. I am thrilled to have had the chance to work with Damon!

Thanks also to Cameron McCarthy and Angharad N. Valdivia who saw enough potential in this book project to accept it for their series "Intersections in Communications and Culture: Global Approaches and Transdisciplinary Perspectives." I am humbled to know that this volume will be published in such distinguished company.

Finally, I am sure I speak for all of the contributors when I thank the girls and young women whose voices speak so loudly and clearly throughout this book. It is their willingness to share their stories, whether through their Internet postings, Web sites, IM conversations, emails, and/or interviews, that gives this book its power and depth. Thank you all.

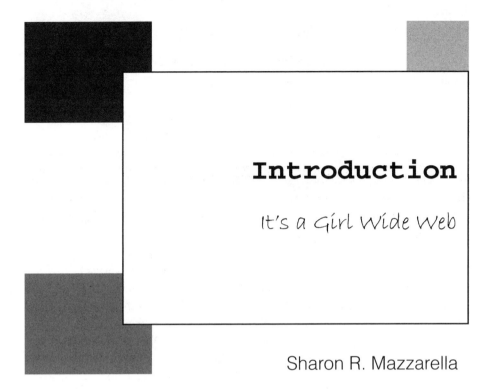

Introduction

It's a Girl Wide Web

Sharon R. Mazzarella

A Growing Presence: Teens Online

In a recent study, the Center for Media Education declared teenagers to be "the *defining users* of [the] digital media culture" who are "as comfortable growing up with digital media as their parents' generation was with the telephone and TV" (Montgomery, 2001, p. 1). In fact, that study reported that almost three quarters of young people aged twelve to seventeen are online. This statistic represents approximately seventeen million young people in this age cohort (Lenhart, Rainie, & Lewis, 2001).

Yet it is not simply the sheer number of teenagers online that is noteworthy, but also what they are doing while there. A recent study supported by the Pew Internet & American Life Project found that the most common uses of the Internet by teens were email (92%), surfing the Web "for fun" (84%), visiting entertainment Web sites (83%), instant messaging[1] (74%), surfing for information on hobbies (69%), news access (68%), playing or downloading games (66%), researching a product/service prior to purchase (66%), listening to online music (59%), visiting chat rooms (55%), and downloading music (53%) (Lenhart et al., 2001). In addition, 24% of online teens have constructed their own personal homepages (Lenhart et al.,

2001).[2] The complete list can be found in the report (Lenhart et al., 2001). According to the study:

> Many American youth say that Internet communication, especially instant messaging, has become an essential feature of their social lives. For them, face-to-face interaction and some telephone conversations have been partially replaced with email and instant message communication. (Lenhart et al., 2001, p. 16)

In his best-selling book *Growing Up Digital: The Rise of the Net Generation*, Don Tapscott argued "there is no issue more important to parents, teachers, policymakers, marketers, business leaders, and social activists than understanding what this younger generation intends to do with its digital expertise" (1998, p. 2).

Contrary to what many in the general public believe, it is not only boys of this generation who are wired. In an article published in *Critical Studies in Mass Communication*, Barbara Warnick pointed out that starting "during late 1996 and 1997, the Web opened up to women," a fact she attributed to the growing number of noncommercial sites created by and for teenage girls and young women (1999, The Web's Changing Nature section, ¶ 1). An August 2000 study reported that girls between the ages of twelve to seventeen were then considered to be the fastest growing group of Internet users (Rickert & Sacharow, 2000). In fact, women's presence online had, by then, surpassed that of men (Rickert & Sacharow, 2000), prompting one author to declare 2000 to be "the year of the female" on the Internet (Rayman-Read, 2001, p. 8). While we often hear stories of the Internet as being a dangerous place for girls or that girls are not as Internet-savvy as boys, the gender gap, indeed, appears to be lessening. Yet research shows "there are significant differences between how boys and girls use the Internet" (Lenhart et al., 2001, p. 6). Girls, for instance are more likely to use email and instant messaging than are boys, while boys are more likely to play/download games and download music (Lenhart et al., 2001). Indeed, such findings are supported by a study of Danish youth in this "new media landscape" (Tufte, 2003, p. 71). The author, Birgitte Tufte, likened this gender difference to the "best friend culture" (2003, p. 72) in which girls tend to engage. Specifically, her research found "that boys tend to play computer games more than girls, whereas girls—if they use computers and have internet access—use their computers to chat" (2003, p. 72).

Despite this growing presence, some scholars have documented the perpetuation of traditional gender roles both in girls' surfing and creating online. For example, Meenakshi Durham's (2001) recent study on lower-income, minority, urban teens' first experience using the Internet, exposed the application of traditional gender roles in their Internet use. Specifically, Durham concluded:

> it was clear that the pre-existing social, and more specifically gender-related, mores and norms of adolescent society governed the ways in which they used the information available on the

Web. From a virtually limitless trove of information, girls and boys selected Web sites that sustained the gender beliefs underpinning "successful" social behaviors in their peer groups. (2001, Conclusions section, ¶ 4)

In response to the growing presence of girls online, there has been a phenomenal growth in academic and popular writings about girls and the Internet. A quick perusal of Amazon.com reveals the existence of a wealth of popular manuals and guides providing advice to girls (or their parents) on using the Internet. With titles such as *Infogirl: A Girl's Guide to the Internet* (Brown, 1999), *Girl Net: A Girls' Guide to the Internet and More!* (Manning, 2001), *Going to the Net: A Girl's Guide to Cyberspace* (Salzman, 1996), and *Net Chick: A Smart-Girl Guide to the Wired World* (Sinclair, 1995), these books evidence the special attention that needs to be paid to bridging the gendered digital divide to make sure we are, as the subtitle of one book put it, "preserving our daughters' place in the cyber revolution" (Furger, 1998). Indeed, one can even find such guides online, including *Internet for Girls: World Wide Web Resource List* which describes itself as a list containing "resources in mathematics and science, parent resources, teacher resources, and sites created especially for women and girls" (Woodka, 1998–2001, ¶ 1). What these books and this Web site have in common is that they are *not* academic publications. Rather, they are guides and/or other advice manuals related to girls' navigation of the Internet, and are targeted to a popular audience.

Similarly, academic scholarship on girls and the Internet is growing. Recent studies, in fact, have begun to focus on girls' active use of the Internet, in particular how the Web sites they create function as forms of identity expression, self-disclosure, and communication (Stern, 1999, 2002a, 2002b; Takayoshi, Huot, & Huot, 1999). The Web recently has been described as "a clubhouse for girls" (Takayoshi et al., 1999) and a place where girls "congregate. . . to rant, rave, and review" (Stapinski, 1999).

Studying Ophelia:
Adolescent Girls, Identity, and Media

Throughout much of the twentieth century, the need for specialized academic and clinical studies of women and girls has been devalued, trivialized and/or marginalized (Brown & Gilligan, 1992; de Beauvoir, 1949; Gilligan, 1982). Not surprisingly, for example, the "defining" research on youth and adolescent development during much of the twentieth century (e.g., Jean Piaget and Lawrence Kohlberg) has told the story of the development of boys (Gilligan, 1982). Indeed, the most often cited scholar of identity development in youth, Erik Erikson, centered his theories on boys. Beginning with the publication of Carol Gilligan's groundbreaking book *In a Different Voice* (1982), a new wave of feminist scholars turned their attention to the unique developmental experiences of girls and women. Gilligan and her colleagues have since authored an influential and oft-cited body of work examining the unique identity development issues, learning strategies, and communicative practices of girls and women.[3] Influencing a new generation of scholars across a range of disciplines—education, psychology, communication, and beyond—this work has played a significant role in the academy's acknowledgment of the importance of studying girls' lives.

While Gilligan's work proved influential in the academic disciplines most aptly labeled as the "social sciences," it was the work of Angela McRobbie which, in the 1970s, first drew the attention of cultural studies scholars to the unique culture of girls (1991).[4] At that time, McRobbie pointed out that the most influential studies of youth culture (e.g., Paul Willis and Dick Hebdige) were studies of *boys'* culture. McRobbie's own work opened the doors for a new generation of cultural studies scholars focused on girls' culture (e.g., McRobbie, 1982, 1984, 1991, 1999).

Beginning with the work of Gilligan and McRobbie, and spurred on by the now-controversial claims of Mary Pipher (1994), feminist scholars across a range of disciplines have taken up the study of what it means to grow up a female in the late twentieth and early twenty-first centuries. In her best-selling book *Reviving Ophelia* (1994), Pipher claims that ours is a "girl poisoning culture" (p. 12) which "smack(s) girls on the head in early adolescence" (p. 23). Based on the findings of her own clinical practice, Pipher asserts that her clients in particular, and girls in general, "are coming of age in a more dangerous, sexualized and media-saturated culture. They face incredible pressures to be beautiful and sophisticated, which in junior high means using chemicals and being sexual" (p. 12). Indeed, the 1990s witnessed a flurry of high-profile studies of adolescent girls' development, many bemoaning the perceived state of crisis of girls in the twentieth-century United States in particular in terms of school performance, self-esteem, and body image.[5] Not surprisingly, the mass media picked up on many of these studies, and began to blast headlines such as: "Perils of Puberty: Girls 'Crash and Burn' in Adolescence" (Eicher, 1994); "Crossing 'Confidence Gap' Poses High Hurdle for Girls (Brecher, 1994); "Suffering in Silence" (Suffering, 1998); "A Perilous Age for Girls" (Mann, 1997); and "Teen Girls No Longer Enjoy an Age of Innocence" (Manning, 1997). The negativity is not limited to the headlines. Through their empirical analysis of newspaper coverage of adolescent girls' lives, Mazzarella & Pecora (2002) concluded that: "While not scapegoating girls for broader social problems, it is clear that the press constructs girls *themselves* as a social problem" (p. 16).

Moreover, a wealth of academic studies have documented the potentially harmful messages—primarily those highlighting the link between a girl's identity and her physical appearance—targeted to girls through the mainstream media. These studies have documented the manner in which such mediated communication, particularly magazines and romance novels, provide girls with images of a stereotypical and narrow range of role models of femininity centered on the importance of beauty, romance, and fashion.[6] Certainly, it cannot be denied that *some* girls rely on this content to *some* extent and that *some* are negatively affected by it.[7]

More recently, the public's concern about girls (as well as boys) and media has centered on the Internet. A recent study, for example, revealed that 57% of parents are concerned that their children will be contacted by strangers online (Lenhart et al., 2001). Despite the fact that nearly 60% of teens report they have been contacted by a stranger online, teens themselves are unconcerned. Fifty-two percent of teens who are active online say they do not worry about this, and only 23% report any "notable level of concern" (Lenhart et al., 2001, p. 5). On the other hand, in keeping with parental fears, the news media perpetuate the idea that the Internet is a dangerous place for youth, in particular for girls. A quick glance at the headlines reveals this concern: "Houston man held in sexual assaults; He met 2 girls over Internet, police say" (Houston man, 2004); "Man, 25, is booked in Internet sex sting; He planned to prey on girl, 14, cops say" (Jensen, 2003); and "Girls lured via Internet" (Girls lured, 2003). (See Lynne Y. Edwards' chapter in this book for a thorough deconstruction of newspaper coverage of girls and the Internet.) Moreover, the academy has not missed out on addressing the concerns over youth and Internet safety, in particular by framing it as a public health issue. (See, for example, Mitchell, Finkelhor & Wolak, 2001; Stahl & Fritz, 2002). Yet, to paraphrase Henry Jenkins (1999), when we spend too much time worrying what the Internet is potentially doing *to* our daughters, we lose sight of what our daughters are doing *with* the Internet.

Girl Wide Web: Girls, the Internet, and the Negotiation of Identity

Scholars of adolescent girls' identity development have pointed out the need for girls to have outlets for self-expression to aid in this development (Brown & Gilligan, 1992; Taylor et al., 1995). (See Shayla Marie Thiel's chapter in this volume for an extensive review of the identity development literature.) Moreover, girls need to feel such outlets are safe. Indeed, concluding the anthology *Growing up Girls: Popular Culture and the Construction of Identity*, Mary Bentley (1999) argued for the need for girls to have safe spaces in their lives. Specifically, she asserted that girls "need spaces where they can know what they know and try new identities without self-censoring. Without safe spaces, girls will not be fully able to discover who they are and who they would like to become" (pp. 219–220). While many in the general public and in the media see the Internet as a potentially *unsafe* space for youth in general, and girls in particular, others see the Internet as offering safe places (Furger, 1998; Hird, 2000)—places where girls "are free to speak without fear of endangering themselves, their relationships, or their 'realities'" (Stern, 1999, p. 39).

Given the prevalence of this medium in the lives of young people in general, and the ever-growing presence of girls online, serious scholarly study of girls and the Internet is imperative. Informed by a range of theoretical perspectives, primarily those within the broader field of cultural studies, and employing a range of methodologies including in-depth interviews, quantitative and qualitative content analysis, surveys, and participant observation, the chapters in this book address a range of issues related to girls and the Internet including commercially produced Web sites targeting pre-adolescent and adolescent girls, Web sites created by girls themselves, girls' postings on newsgroups, the creation of online communities for girls, the increasing popularity of instant messaging, as well as newspaper coverage of girls and the Internet. What they all share, however, is an emphasis on the link between the Internet and identity development.

The book begins by confronting the public's fear of the Internet as a dangerous place for girls. Grounded in frame theory, Lynne Y. Edwards' chapter "Victims, Villains, and Vixens: Teen Girls and Internet Crime" analyzes major newspaper articles' reporting on girls and Internet crime. Through rigorous content analysis, Edwards documents journalists' over-reliance on law enforcement sources and concurrent failure to include the voices of girls themselves. As a result, these articles primarily frame the Internet as a dangerous place for girls, in particular as a result of girls' potential to be victimized by other Internet users.

While adults are concerned about potentially dangerous strangers lurking in cyberspace, they also are concerned about what other types of content young people will encounter when surfing the Net, in particular when it comes to commercially produced Web sites. Coming from the opposite perspective that the Web might, indeed, provide *safe* spaces for girls to speak, Ashley D. Grisso and David Weiss, in their chapter "What are gURLs Talking about? Adolescent Girls' Construction of Sexual Identity on gURL.com," listen to the "voices" of girls who post messages on the commercial site gURL.com. Specifically focusing on girls' construction of sexual identity through their postings, Grisso and Weiss document how some girls use this safe space to perform, negotiate, and construct their sexual identities.

Examining another potentially girl-empowering commercial site, Debra Merskin deconstructs About-Face.com, a site grounded in the principles of media literacy, and dedicated to exposing the myths, stereotypes, and strategies in advertising—in particular advertising targeted to women and girls. In her chapter "Making an About-Face: Jammer Girls and the World Wide Web," Merskin exposes the "false dichotomy" often plaguing girls' identity development—the idea that one can either be a good girl or a bad girl—and creates the category of females she calls "Jammer Girls"—girls and young women who are critical of popular culture messages and who actively work to subvert such messages both for themselves and for others.

With the growing popularity of the Internet, it was inevitable that previously print-only publications would produce online versions or even migrate solely to the Internet. Traditional magazines for teen girls including *Seventeen*, *YM*, and *Teen People* have joined the crowded roster of magazines creating companion online versions. Yet the Internet also has opened up space (literally and figuratively) for alternatives to these traditional magazines to reach girls, and it is one such magazine, Blue Jean Online, that Susan F. Walsh analyzes in her chapter "Gender, Power, and Social Interaction: How Blue Jean Online Constructs Adolescent Girlhood." Her feminist analysis shows how this online alternative presents girls with a broader and more diverse definition of femininity than its more traditional print and online counterparts.

Magazines are not the only "traditional" medium to find a presence on the Web. As Seiter (2002, p. 35) makes clear, as the "resemblances between websites and television programming increase," the more the Web becomes mainstream. Indeed, in their chapter, "Exploring Dora: Re-embodied Latinidad on the Web," Susan J. Harewood and Angharad N. Valdivia interrogate the Web presence of Nickelodeon's animated *Dora the Explorer* children's program. Specifically, by focusing on parental postings on *Dora's* message board, they expose parents' role in constructing and negotiating Dora's Latina identity through application of such tropes as the tropics, the Spanish language, and the body.

Usenet newsgroups were one of the first venues offering adolescent girls the opportunity to "speak" online. Christine Scodari's qualitative content analysis of adolescent girls' posting to various television Usenet newsgroups documents the manner in which girls use this opportunity to speak as a way to affirm their senses of self and place. At the same time, however, Scodari's chapter, "You're Sixteen, You're Dutiful, You're Online: 'Fangirls' and the Negotiation of Age and/or Gender Subjectivities in TV Newsgroups," finds that when girls are provided an outlet for their voices, they often replicate the same codes of romantic individualism first identified by McRobbie some thirty years ago.

Following Susannah Stern's lead, and heeding Mary Celeste Kearney's call for more analysis of girls as producers of mediated content, Kimberly S. Gregson examines Web sites created by young, female fans of *shoujo* anime—defined as a sub-genre of Japanese animation (anime) featuring strong female lead characters and targeted primarily to female viewers. In "What if the Lead Character Looks Like Me? Girl fans of *Shoujo* Anime and Their Web Sites," Gregson finds, surprisingly, that these young female fans do not appear to be identifying with *shoujo*'s strong female lead characters. Informed by Angela McRobbie, Gregson exposes fans' tendency to celebrate the male characters and the ensuing romantic subplots of *shoujo*.

Informed both by McRobbie's (1991) theories of "romantic individualism" and John Fiske's (1992) "cultural economy of fandom," Sharon R. Mazzarella, in her chapter, "Claiming a Space: The Cultural Economy of Teen Girl Fandom on the Web, " deconstructs Web sites created by female fans of teen-oriented actor Chad Michael Murray. On the surface, she finds these sites tend to replicate traditional, mass-produced teen idol magazines, but she argues that focusing on that would be to miss the more important components of these sites. Acknowledging the role of girls as cultural producers, in this case of both a technologically complex product and of a community of like-minded fans, Mazzarella shows that girls use their technological expertise to create a safe space for themselves and other girls to engage in an often denigrated cultural activity—teen idol fandom.

Divya C. McMillin, in her chapter, "Teen Crossings: Emerging Cyberpublics in India," engages in ethnographic analysis of the use of the Internet and other media by teenage girls in Bangalore, India. Her findings reveal a complex weaving of new technologies within traditional gendered practices (for example, the expectation that girls' leisure will be primarily home-based), facilitating girls' (virtual) boundary-crossings. Moreover, she argues that, while only a small part of girls' leisure activities, email and surfing the Web, are, however, essential components of what she calls "a matrix of rituals of identity expression" for these girls.

The previously mentioned Pew Internet and American Life study found that while there are "negligible differences" between teens' and adults' use of the Internet for some things like email, when it comes to instant messaging, the difference are much more noticeable. Seventy-four percent of online teens report using instant messaging while a companion study found that only 44% of adults did so (Lenhart et al., 2001). It is this generational phenomenon that Shayla Marie Thiel examines in her chapter, "'IM Me': Identity Construction and Gender Negotiation in the World of Adolescent Girls and Instant Messaging." Specifically, through examining the "cultural narratives" common across girls' interviews and IM "conversations," Thiel documents the manner in which IM facilitates girls' negotiation and articulation of their identities, in particular their gender identities.

Examining a range of new media use including instant messaging, Lynn Schofield Clark provides evidence of a phenomenon she labels "constant contact"—a situation facilitated by these technologies, which enable young people to be able to contact their peers at a moment's notice or even to remain in continuous contact. Informed by in-depth interviews, Clark argues in her chapter, "The Constant Contact Generation: Exploring Teen Friendship Networks Online," that teenage girls' use of the Internet and other new media provides them with a form of control over their environment—both of their parents' supervision of them and their presentation of self within the peer community. Clark concludes by pondering the risks and benefits related to a generation so continuously in contact with one another.

Taken together, these chapters provide a rich portrait of the complex relationship between girls, the Internet, and the negotiation of identity—a relationship often linked with the creation of safe spaces and community.

Notes

1. Instant Messaging (IM) is a technology that allows computer users to have real-time, text-based "conversations" with others online. All it takes is either a computer with an Internet connection, a hand-held PDA, or a cell phone with an LCD screen as well as free downloadable software from services such as Yahoo!, America Online, or MSN. IM users create "buddy lists" containing the screen names of others with whom they wish to engage in online conversations.
2. Personal homepages, as defined by Susannah Stern (2002b) are "individually generated WWW sites whose focus is predominantly personal" (Background section, ¶ 2). Chandler and Roberts-Young liken teenagers' personal homepages to "the environmental space of teenagers' own bedrooms in their real-life homes" (1998, p. 3).
3. See, for example, Brown & Gilligan, 1992; Gilligan, Lyons & Hanmer, 1990; Gilligan, Rogers, & Tolman, 1991; and Taylor, Gilligan & Sullivan, 1995.
4. This source contains reprints of McRobbie's work spanning a thirteen-year period beginning in 1978.
5. The most notable of these include: American Association of University Women, 1991; Brown and Gilligan, 1992; Brumberg, 1997; Orenstein, 1994; Sadker and Sadker, 1994.
6. See, for example, Christian-Smith, 1988, 1990; Duffy & Gotcher, 1996; Durham, 1998; Evans, Rutberg, Sather & Turner, 1991; Guillen & Barr, 1994; Mazzarella, 1999; McRobbie, 1991; Pecora, 1999; Peirce, 1990, 1993.
7. See, for example, Duke & Kreshel, 1998; Durham, 1999; Levine, Smolak, & Hayden, 1994; Martin & Gentry, 1997; Martin & Kennedy, 1993; Turner, Hamilton, Jacobs, Angood, & Dwyer, 1997.

References

American Association of University Women. (1991). *Shortchanging girls, shortchanging America: A nationwide poll to assess self-esteem, educational experiences, interest in math and science, and career aspirations of girls and boys aged 9–15.* Washington, D.C.: Author. ED 340–657.

Bentley, M. K. (1999). The body of evidence: Dangerous intersections between development and culture in the lives of adolescent girls. In S. R. Mazzarella, & N. Pecora (Eds.), *Growing up girls: Popular culture and the construction of identity* (pp. 209–221). New York: Peter Lang.

Brecher, E. J. (1994, September 30). Crossing "confidence gap" poses high hurdle for girls [Electronic version]. *Miami Herald*, p. F1.

Brown, M. (1999). *Infogirl: A girl's guide to the Internet.* New York: Rosen Publishing.

Brown, L. M., & Gilligan, C. (1992). *Meeting at the crossroads: Women's psychology and girls' development.* Cambridge, MA: Harvard University Press.

Brumberg, J. J. (1997). *The body project: An intimate history of American girls.* New York: Random House.

Chandler, D., & Roberts-Young, D. (1998). The construction of identity in the personal homepages of adolescents. Retrieved March 17, 2004, from http://www.aber.ac.uk/media/Documents/short/strasbourg.html

Christian-Smith, L. K. (1988) Romancing the girl: Adolescent romance novels and the construction of femininity. In L. G. Roman, L. K. Christian-Smith, & E. Ellsworth (Eds.), *Becoming feminine: The politics of popular culture* (pp. 76–101). London: Falmer Press.

Christian-Smith, L. K. (1990). *Becoming a woman through romance.* New York: Routledge.

de Beauvoir, S. (1949). *The second sex.* New York: Knopf.

Duffy, M., & Gotcher, J. M. (1996). Crucial advice on how to get the guy: The rhetorical vision of power and seduction in the teen magazine *YM. Journal of Communication Inquiry, 20,* 32–48.

Duke, L. L., & Kreshel, P. J. (1998). Negotiating femininity: Girls in early adolescence read teen magazines. *Journal of Communication Inquiry, 22,* 48–71.

Durham, M. G. (1998). Dilemmas of desire: Representations of adolescent sexuality in two teen magazines. *Youth and Society, 29,* 369–389.

Durham, M. G. (1999). Girls, media, and the negotiation of sexuality: A study of race, class, and gender in adolescent peer groups. *Journalism & Mass Communication Quarterly, 76,* 193–216.

Durham, M. G. (2001). Adolescents, the Internet and the politics of gender: A feminist case analysis [Electronic version]. *Race, Gender & Class, 8*(4), 20–41.

Eicher, D. (1994, July 5). Perils of puberty: Girls "crash and burn" in adolescence [Electronic version]. *Denver Post,* p. E01.

Evans, E. D., Rutberg, J., Sather, C., & Turner, C. (1991). Content analysis of contemporary teen magazines for adolescent females. *Youth and Society, 23,* 99–120.

Fiske, J. (1992). The cultural economy of fandom. In L. Lewis (Ed.), *Adoring audience: Fan culture and popular media* (pp. 30–49). New York: Routledge.

Furger, R. (1998). *Does Jane compute? Preserving our daughters' place in the cyber revolution.* New York: Warner Books.

Gilligan, C. (1982). *In a different voice: Psychological theory and women's development.* Cambridge, MA: Harvard University Press.

Gilligan, C., Lyons, N., & Hanmer, T. (Eds.). (1990). *Making connections: The relational world of adolescent girls at the Emma Willard School.* Cambridge, MA: Harvard University Press.

Gilligan, C., Rogers, A. G., & Tolman, D. L. (Eds.). (1991). *Women, girls & psychotherapy: Reframing resistance.* New York: Harrington Park Press.

Girls lured via Internet [Electronic version]. (2003, September 27). *The Advertiser,* p. 42.

Guillen, E. O., & Barr, S. I. (1994). Nutrition, dieting, and fitness messages in a magazine for adolescent women, 1970–1990. *Journal of Adolescent Health, 15,* 464–472.

Hird, A. (2000). *Learning from cyber-savvy students: How Internet-age kids impact classroom teaching.* Herndon, VA: Stylus Publishing.

Houston man held in sexual assaults; He met 2 girls over Internet, police say [Electronic version]. (2004, February 07). *The Houston Chronicle,* p. 37.

Jenkins, H. (1999). *Congressional testimony on media violence.* Retrieved July 16, 2001, from http://media-in-transition.mit.edu/articles/dc.html

Jensen, L. (2003, October 18). Man, 25, is booked in Internet sex sting; He planned to prey on girl, 14, cops say [Electronic version]. *Times-Picayune,* p. 1.

Kearney, M.C. (1998). Producing girls. In S. A. Inness (Ed.), *Delinquents & debutantes: Twentieth-century American girls' cultures* (pp. 285–310). New York: NYU Press.

Lenhart, A., Rainie, L., & Lewis, O. (2001). Teenage life online: The rise of the instant-message generation and the Internet's impact on friendships and family relationships. Washington, DC: Pew Internet and American Life Project. Retrieved March 17, 2004, from http://www.pewinternet.org/

Levine, M. P., Smolak, L., & Hayden, H. (1994). The relation of sociocultural factors to eating attitudes and behaviors among middle school girls. *Journal of Early Adolescence, 14*, 471–490.

Mann, J. (1997, October 10). A perilous age for girls [Electronic version]. *Washington Post*, p. E03.

Manning, A. (1997, October 6). Teen girls no longer enjoy an age of innocence [Electronic version]. *USA Today*, p. 4D.

Manning, S. (2001). *Girl Net: A girls' guide to the Internet and more!* New York: The Chicken House.

Martin, M. C., & Gentry, J. W. (1997). Stuck in the model trap: The effects of beautiful models in ads on female pre-adolescents and adolescents. *Journal of Advertising, 26*, 19–33.

Martin, M. C., & Kennedy, P. F. (1993). Advertising and social comparison: Consequences for female preadolescents and adolescents. *Psychology and Marketing, 10*, 513–530.

Mazzarella, S. R. (1999). "The Superbowl of all dates": Teenage girl magazines and the commodification of the perfect prom. In S. R. Mazzarella, & N. O. Pecora (Eds.), *Growing up girls: Popular culture and the construction of identity* (pp. 97–112). New York: Peter Lang.

Mazzarella, S. R., & Pecora, N. (2002, November). Girls in crisis: Newspaper coverage of adolescent girls. Paper presented at the meeting of the National Communication Association, New Orleans, LA.

McRobbie, A. (1982). *Feminism for girls*. London: Routledge.

McRobbie, A. (1991). *Feminism and youth culture: From* Jackie *to* Just Seventeen. Boston: Unwin Hyman.

McRobbie, A. (1999). *In the culture society: Art, fashion, and popular music*. London: Routledge.

McRobbie, A., & Nava, M. (Eds.). (1984). *Gender and generation*. London: Macmillan.

Mitchell, K., Finkelhor, D., & Wolak, J. (2001). Risk factors for and impact of online sexual solicitation of youth [Electronic version]. *JAMA, 285*, 3011–3014.

Montgomery, K. C. (2001). *Teensites.com: A field guide to the new digital landscape*. Washington, DC: Center for Media Education. Retrieved July 10, 2003, from http://www.cme.org/teenstudy/

Orenstein, P. (1994). *Schoolgirls: Young women, self-esteem, and the confidence gap*. New York: Doubleday.

Pecora, N. (1999). Identity by design: The corporate construction of teen romance novels. In S. R. Mazzarella, & N. O. Pecora (Eds.), *Growing up girls: Popular culture and the construction of identity* (pp. 49–86). New York: Peter Lang.

Peirce, K. (1990). A feminist theoretical perspective on the socialization of teenage girls through *Seventeen* magazine. *Sex Roles, 23*, 491–500.

Peirce, K. (1993). Socialization of teenage girls through teen-magazine fiction: The making of a new woman or an old lady? *Sex Roles, 29*, 59–68.

Pipher, M. (1994). *Reviving Ophelia: Saving the selves of adolescent girls*. New York: Ballantine Books.

Rayman-Read, A. (2001, January 29). Gurl power [Electronic version]. *The American Prospect*, p. 8.

Rickert, A., & Sacharow, A. (2000). *It's a woman's World Wide Web*. Media Matrix & Jupiter Communications. Retrieved July 7, 2003, from http://pj.dowling.home.att.net/4300/women.pdf

Sadker, M., & Sadker, D. (1994). *Failing at fairness: How America's schools cheat girls*. New York: Charles Scribner's Sons.

Salzman, M. (1996). *Going to the Net: A girl's guide to cyberspace*. New York: Avon Books.

Seiter, E. (2002). New technologies. In T. Miller (Ed.), *Television studies* (pp. 34–37). London: BFI Publishing.

Sinclair, C. (1995). *Net chick: A smart-girl guide to the wired world*. New York: Holtzbrinck Publishers.

Stahl, C. & Fritz, N. (2002). Internet safety: Adolescents' self-reports [Electronic version]. *Journal of Adolescent Health, 31*(1), 7–10.

Stapinski, H. (1999). What works—where the girls are: Women 18-to-34 congregate at the Girls on Web site to rant, rave, and review. *American demographics, 21*(1), 47–52.

Stern, S. R. (1999). Adolescent girls' expression on web home pages: Spirited, sombre and self-conscious sites. *Convergence: The Journal of Research into New Media Technologies, 5*(4), 22–41.

Stern, S. R. (2002a). Sexual selves on the World Wide Web: Adolescent girls' home pages as sites for sexual self-expression. In J. D. Brown, J. R. Steele, & K. Walsh-Childers (Eds.), (2002). *Sexual teens, sexual media: Investigating media's influence on adolescent sexuality* (pp. 265–285). Mahwah, NJ: Lawrence Erlbaum.

Stern, S. R. (2002b). Virtually speaking: Girls' self-disclosure on the WWW [Electronic version]. *Women's Studies in Communication, 25*(2), 223–252.

Suffering in silence [Electronic version]. (1998, March 31). *Washington Post*, p. Z05.

Takayoshi, P., Huot, E., & Huot, M. (1999). No boys allowed: The World Wide Web as clubhouse for girls. *Computers and Composition, 16*, 89–106.

Tapscott, D. (1998). *Growing up digital: The rise of the Net generation*. New York: McGraw-Hill.

Taylor, J. M., Gilligan, C., & Sullivan, A. M. (1995). *Between voice and silence: Women and girls, race and relationship*. Cambridge, MA: Harvard University Press.

Tufte, B. (2003). Girls in the new media landscape [Electronic version]. *Nordicom Review, 24*(1), 71–78.

Turner, S. L., Hamilton, H., Jacobs, M., Angood, L. M., & Dwyer, D. H. (1997). The influence of fashion magazines on the body image satisfaction of college women: An exploratory analysis. *Adolescence, 32*, 603–614.

Warnick, B. (1999) Masculinizing the feminine: Inviting women on line ca. 1997 [Electronic version]. *Critical Studies in Mass Communication, 16*(1), 1–19.

Woodka, D. (1998–2001). *Internet for girls: World Wide Web resource list*. Retrieved March 17, 2004, from http://www.sdsc.edu/~woodka/resources.html

1

Victims, Villains, and Vixens

Teen Girls and Internet Crime

Lynne Y. Edwards

Introduction

The Internet is a great place for girls to find recipes for cooking up . . . drugs. Or to make friends in a chat room . . . with sexual predators. Or to find really cool images of . . . pornography. Certainly, the public's perception is that the Internet can be a scary place for youth, a perception reinforced by statistics showing that one in five youths surveyed were approached or solicited sexually online and 25% received unsolicited pornography (Finkelhor, Mitchell, & Wolak, 2000). But the perception is even more pronounced when it comes to the Internet dangers that potentially could befall girls—at least this is what major U.S. newspapers would have us believe when they tell us about the magnitude of Internet crimes against girls, the few Internet crimes committed by girls, and the heroic efforts by police to save them. This ability of the news media to influence our perceptions about social issues and our options regarding solutions is a component of the surveillance function the news plays in our lives—a function that is validated by its reliance on quotes from official sources. Through highlighting the voices and agendas of official sources, the news media tell us what is going right and what is going wrong in our world—what to worry about and what to celebrate. But what are the news media *really* telling us about girls and Internet crime? Are girls at risk every time they log on to their computers? Are *we* at risk every time they do? On whom can we depend to "pro-tect" us . . . and at what cost?

This chapter explores these questions by examining the frames employed by the news media when reporting on Internet crimes committed by and against girls. I argue that news frames construct: 1) girls as victims of the Internet, 2) police as the *only* heroes who can save them, and 3) our civil liberties as the ultimate sacrifice we may have to make in this battle to protect girls.

Structure of News

News as symbolic process

News is not reality, but a representation of reality. It does not tell it "like it is," but rather, "like it means" (Bird & Dardenne, 1988, p. 71). As a representation of real-ity, news is metalanguage; its meaning and, therefore, the meaning of our reality come from the specific arrangement of its elements (Barthes, 1982). From the tra-ditional inverted pyramid and narrative forms to the "facts" of "who, what, where, when, why and how" as related by "sources," news is a unique symbolic system that has the power to legitimate for consumers the very symbols it presents (Allan, 1999). It is the ongoing, daily repetition of this metalanguage that allows news to sustain its cultural power. As our cultural scribes and repositories of knowledge, journalists and the institution of the news media play a role in defining social and political dis-course and, by extension, our social and political reality. As Bernard Cohen (1963) so eloquently stated:

> The press is significantly more than a purveyor of information and opinion. It may not be successful much of the time in telling people what to think, but it is stunningly successful in telling readers what to think about. (p. 13)

The press accomplishes this through a unique symbolic process by which news is culturally empowered to serve a surveillance function in our society. This function allows news producers to select *what* information we receive, from *whom* we receive it, and *how* we receive it. Wright (1986) suggests this function carries several social consequences. First, news reaffirms our social order by defining our villains, victims and heroes, and heroes maintain social order by punishing villains who threaten it by violating social norms. Second, status is conferred to the issues and individuals featured in news. Third, news can heighten anxieties by extensive or sensational cov-erage, especially when there is no interpretative or mediating information provid-ed. In other words, news gives us the impression that we can understand and control our world by reaffirming the values of our social and political order as well as implicitly teaching us which behaviors are rewarded and which are punished. In so doing, ordinary people are elevated in importance by coverage which suggests that sources, willing or otherwise, are powerful forces in the construction of news.

As our only gateway to this information, journalists are expected to be unbiased and objective observers. In the interest of objectivity, journalists are prohibited from having or using their voices to tell the story (Lipari, 1996). Although these values are presented to us *by* journalists, they often originate *with* their sources whose values are implicit in the information they provide (Gans, 1979; Koch, 1990; Zoch & Turk, 1998). Suspects are vilified and heroes (as well as, at times, victims) are reified through their roles not only in the news story but also as news sources. As witnesses, experts, victims and suspects, these sources are defined as important by their presence in the news and by their unique position to tell their stories. Eyewitnesses, authorities, victims and experts provide information; their names and narratives put a human face on the facts and the stats. Their presence in news stories not only elevates the status of these sources but also the status of the journalist. First, citing others demonstrates the journalist's access to information and to the people who have it, which legitimates the journalist's role as repository of cultural information. Second, selecting and citing sources demonstrate the journalist's power to legitimate the source and to deny that legitimation to other sources (Ericson, Baranek & Chan, 1991).

The power to legitimate self and source is an area of concern, however. The frequent selection of official sources marginalizes ordinary people as potential sources, and the people most often marginalized are youth. Sadly, the sources frequently missing from news stories about youth and crime are youths themselves, an omission that leads to their frequent misrepresentation in news (Males, 1994). Rather than including youth as sources in stories about them, law enforcement officials and experts are often cited, providing explanations for why youths commit crime, get pregnant, or any number of other issues associated with them (Mazzarella & Pecora, 2002; Shepard, 2002). Even when youth do appear in news stories about juvenile crime, the predominant speakers are "white male adults" (Dorfman & Woodruff, 1998, p. 84). This absence of youth as sources in news stories is particularly ironic at a time when journalists are beginning to identify juvenile suspects by name (Shepard, 1994).

Framing as symbolic process

"Framing" is the process of organizing information in a way that provides a particular interpretation or meaning for the audience (Gamson & Modigliani, 1989; Iyengar, 1991; Neuman, Just & Crigler, 1992; Tuchman, 1978). Frames develop from the processes of "cultural resonance, sponsor activities, and media practices" (Gamson & Modigliani, 1989, p. 5). Cultural resonances are larger social themes, such as youth violence, that are familiar and meaningful to audiences. Sponsor activities are actions by individuals such as law enforcement spokespersons who have their own agendas and who are trying to get their information into the news. Finally, media practices are the news norms and processes involved in producing news, including story selection, sources quoted, editing, and information organization. This imbalance in source access, unfortunately, contributes to the over-reliance on official sources which, in turn, contributes to the predominance of law-and-order frames in news.

One of the most important aspects of the framing process, however, is its perceived effect on readers. The significance of the framing process lies in its ability to limit the audience's perceived options for resolving those issues deemed to be newsworthy. In his book, *Is Anyone Responsible?*, Shanto Iyengar (1991), for example, found that when stories were framed as "episodic" and focused on issues as concrete events, the public tended to blame individuals for their plight. When stories were framed as "thematic" and focused on issues as part of a general problem, the public blamed society for the problem. In a recent analysis of youth-crime stories in three California newspapers, McManus and Dorfman (2002) found that most stories were presented in an "episodic" frame, a finding that suggests readers may blame youths for their violent acts. The authors, however, noted their study was limited by the small number of newspapers in their sample. Children Now (2001) reported that children appear in California local news stories less frequently than their actual population numbers would indicate but are over-represented in crime stories. Stevens (2001), too, reported that news coverage over-represented youth crime and that news coverage of youth is primarily related to violence with few, if any, nonviolent depictions of young people in the news. This imbalanced portrayal, the authors surmise, may lead to misperceptions of young people as dangerous and as a threat to society. One effect of this narrow portrayal is the increased support by the public for punitive crime policies compared to preventive policies. For example, in a study on the effects of framing in political rhetoric, Mackey-Kallis and Hahn (1994) suggest that framing the drug problem as a "war on drugs" may have contributed to our failure to solve the problem by "misdiagnosing" the issue; misdiagnoses lead to ineffective solutions.

Girls and Cybercrime

In order to study this topic, I examined 125 articles[1] about teen girls and Internet crime from January 1, 1990 to January 1, 2002 in five major daily newspapers.[2] Analysis included headlines, leads, sources quoted, and six frame categories: *girls-as-risk*, *girls-at-risk*, *adults-as-risk*, *adults-at-risk*, *police-as-hero*, and *civil liberties-at-*

risk.[3] Only 6 articles (5%, n = 125) focused on the girls as victims or villains. Four articles (3%, n = 125) were categorized in the girls-as-risk frame, while only 2 of the articles (2%, n = 125) were categorized in the girls-at-risk frame; both frames featured police as sources in all articles. In stark contrast to the girl-frames, 40 articles (32%, n = 125) were categorized in the adults-as-risk frame, with attorneys cited as sources in 25 of the 40 articles (63%) and police cited as sources in only 12 (30%, n = 40). In the adults-at-risk frame, however, there were only 7 articles (6%) and they featured police sources cited in 4 articles (57%); however, experts were cited in 5 of the articles (71%).

The largest block of articles, 48 out of 125 (38%) were framed as police-as-hero and, not surprisingly, featured police as sources in 73% (n = 48). Attorneys were cited in 19 articles (40%, n = 48) and experts in 7 articles (15%, n = 48). Finally, 13 of the 125 articles (10%) were framed as civil liberties-at-risk, with police and parents cited as sources in 15% of the articles (n = 13). Experts were cited in 3 (23%) of the articles in this frame, and attorneys were cited in 7 or 54% of the articles in this frame. Given the perceived threat to civil liberties evident in this frame, it seems logical that so many articles would feature attorneys as sources. The remaining 11 of the 125 articles (9%) that did not fit into these six frames were categorized as "other."

Cyber-Villains: Girls-as-risk

The 4 articles in the girls-as-risk frame situate the girls' crimes in female "spaces," suggesting that the use of the technology was merely an extension of stereotypically female behaviors occurring in stereotypically female spaces like the kitchen and the bedroom. More disturbingly, however, the girls appear to be technologically inept, lacking the agency to successfully commit these crimes on their own. Police and law enforcement officials are quoted as sources in all 4 articles (100%, n = 4); the neighbors and an uncle of a youth suspect are quoted in 1 of the articles (25%, n = 4).

In the article "Girl, 15, arrested making drugs at home with Internet recipe" ("Girl, 15, arrested," 1997), the lead describes the crime this way: "Investigators say a 15-year-old girl they caught cooking up narcotics in her home learned the recipe on the Internet." The use of food-preparation metaphors like "cooking up narcotics" and "learned the recipe" implicitly situate this crime in a stereotypically female space (the kitchen), which is reaffirmed later in the article when officers report that "she had already thrown away two failed batches." More importantly, however, official sources in this frame are presented as active and effective in the fight against crime by catching and arresting the girl "on suspicion of manufacturing a controlled substance," a crime that seems far more serious than merely "cooking up" narcotics as described in the lead.

This framing continues in the follow-up article published the next day. In the headline, "California and the West; Internet drug recipe raises officials' fears" (Hayes, 1997), we see that it is not what the girl did with the "recipe" but that it came from the Internet that makes the headline. It is the information, and not the girl's actions, that "frightens" officials. Her lack of power or expertise with the Internet is further suggested a few paragraphs later: "The incident—which comes on the heels of others involving youths who have tangled with danger while toying with the Internet—raises new concerns about the amount of information that can be accessed at the tap of a keyboard." Phrases like "tangled with danger," "toying with the Internet," and "tap of a keyboard" do not suggest a serious recreational-drug manufacturer setting up shop with information downloaded from the Internet, but rather, a girl at play. The article also implicitly raises the female "space" stereotype in the language used to describe the information that was downloaded from the Internet and her lack of agency in obtaining it:

> The teenager, whose name is not being released because of her age, said the step-by-step instructions came from the Internet. But police say the handwritten recipe found in her backpack indicates she may have had help from others trying to conceal the project from law enforcement. Police said the recipe was with a note that said: "To you, from me . . ."

Although the girl refers to the information as "step-by-step instructions," the journalist quotes police as describing the "instructions" as a "recipe." Finally, the journalist ends the article by raising the specter of a troubled home "space" by reporting that the girl was living with her uncle when she was arrested; an uncle who describes his niece as "a bright student who was probably 'just stirring things up' rather than launching a serious drug manufacturing lab." Although the uncle's description of his niece seems to re-affirm his ignorance of her activities, the placement of the quote at the end of the article, following several quotes by law enforcement officials about the prevalence of drug-making information on the Internet, serves to also re-affirm the officers' status as the only heroes who can protect us from the evils of the Internet.

The headline of the article "Teen who had bomb material is arrested; Irvine girl, 16, is in custody after discovery of materials and instructions triggered evacuation of apartment complex" (Roug and Harris, 2000) suggests an emphasis on law-and-order in preventing a crime from occurring, and not the availability of bomb-making instructions online. The phrases "Teen . . . is arrested," "is in custody," "evacuation of apartment complex" all indicate a successful response by police to a potential bomb threat, a threat made possible, we learn in the lead, by the Internet: "A 16-year-old Irvine girl was arrested Friday after police found flammable chemicals and Internet-supplied 'bomb-recipes' stashed in the closet of her bedroom." It is in the lead paragraph that we learn the teen's gender and the source of the bomb-making information—information that is, again, described as "recipes." Similar to the narcotics articles, the troubled home "space" was implicitly raised in this article by reporting that the bomb materials were, unbeknownst to her mother, found "stashed in the closet of her bedroom." Further, the girl was not home when the chemicals were discovered and did not return home until two days later, a fact neither explored nor explained in the article.

In the fourth article, "Police say girl sent AOL threats to self; teen allegedly tried to blame others" (White, 2001), teenage girls' computer crimes are also attributed to the availability of the technology for use in common disputes. The girl in the headline claimed that two other girls in her school were sending her threats via Instant Messenger when, in reality, she sent the messages to herself and faked the evidence to implicate the other girls. The official sources quoted in the article downplay the criminal activity involved and place blame on the technology, but not on the girl who used it to harass her classmates: "Authorities characterized the episode as a high school dispute gone wrong and said it is an example of how the Internet has become a frequent source of crime for youngsters." The girl's crime is further minimized as authorities go on to say: "Teenagers may assume they are anonymous on the Internet, but there are ways of finding out who they are. Frankly, it's not that tough to figure it out, so they shouldn't even be trying." By citing an official source saying it was not "tough to figure it out," the article discounts the intricacy of the crime and the effort expended by the girl to implicate the others while at the same time, re-affirming their ability to restore order.

Cyber-Victims: Girls-@-risk

As in the articles in the girls-as-risk category, the 2 articles in the girls-at-risk category include police as sources, and situate the girls in gendered spaces that reinforce their technological victimization. Unlike the girls-as-risk articles, there is a youth source cited in 1 article in this category. Even with a youth source, however, both articles primarily present the two girls through the words of others. In one article, key portions of the victim's court testimony are presented as they are paraphrased by attorneys; in the other article, the youth is a murder victim whose character is presented through quotes from law enforcement officials, neighbors, and her high school principal. In both articles, the girls' voices are appropriated by adult sources.

The story of Christina Long best exemplifies this framing. Christina Long was a thirteen-year-old girl who was murdered in May 2002 by Saul Dos Reis, a twenty-five-year-old man she met in an Internet chat room. In the lead paragraphs of the *New York Times* article, "Slain girl used Internet to seek sex" (Kilgannon, 2002), she is presented in conflicting imagery:

> By day, she was Christina Long, a 13-year-old altar girl and a co-captain of the cheerlead-
> ing team at St. Peter Roman Catholic School in Danbury, Conn., where the principal said
> she was a "good student and well behaved" . . . But in the evenings, the authorities say, she
> logged onto the Internet using the screen name "LongToohot4u" and the slogan, "I will do
> anything at least once." In her bedroom, the police say, she used her computer to troll chat
> rooms and meet adult men for sex, her marital status listed as, "I might be single I might
> not be."

By capturing an image of the victim as "good student and well-behaved," according to her principal, yet willing "to troll chat rooms and meet adult men for sex,"
the article suggests Christina is to blame for her predicament and that the adults
in her life were unaware of her Internet activities. By situating the victim's behaviors in specific times and spaces through the use of school and police authorities as
sources, this quote also reinforces gender stereotypes and parental fears about the
Internet. The "good" Christina comes out in the morning at school where her conduct is presumably supervised. The "bad" Christina acts out through the Internet
in the evenings at home in her *bedroom*, a space already problematized by other risky
behaviors ascribed to unsupervised girls.

This image is further borne out with information provided by the police: "The
police said Christina used many provocative screen names and routinely had sex with
partners she met in chat rooms. Her online profile includes photos of her that hardly resemble the tomboy image in those released by the authorities." The inclusion
of this information about Christina "routinely having sex" with strangers and the
use of "many provocative screen names" is particularly troubling for two reasons.
First, it suggests that authorities blame the victim for actively seeking out the man
who ultimately killed her; by stating that her "online profile" does not resemble the
"official" tomboy image, the journalist appears to assign blame to Christina. Second,
there is no such information provided about Dos Reis, her alleged murderer. Neither
the police nor the news media provide any information about his sexual or criminal history.

The lead to the November 2002 *New York Times* article "On Stand, Girl
Recalls Week of Captivity and Rape on L.I." by Elissa Gootman, exemplifies the
additional trauma girls experience when they *do* report sexual assaults and pursue
prosecution. We also *see* the presentation of a traumatized victim testifying in open
court, but we do not *hear* her, despite what the headline suggests:

> In a soft, almost babyish voice, a 16-year-old Massachusetts girl testified today about a week-
> long ordeal in which she said a Long Island man who had lured her over the Internet held
> her captive, assaulted her, and ordered her to have sex with him and his girlfriend, bound
> her with rope and choked her until she passed out.

By defining the victim's *voice* as "soft" and "almost babyish," and not allowing
her voice *to tell* her story, the news media negates the use of the youth source in this

story. Although the lead emphasizes her youth and victimization, the absence of a direct quote from the victim denies her agency in her testimony. This problem continues with the inclusion of a direct quote from the victim that indicates her willingness to participate in her own victimization: "'I said I had a lot of problems,' she told the jury. 'He said that if I had sex with him when he wanted, then I could do whatever I wanted outside of that.'" While the victim is allowed to tell us her role in the crimes committed against her, she is not permitted to directly tell us about the horrors she experienced at the hands of her captors or the steps she took to escape them.

Cyber-Villains: Adults-as-risk . . . to the social order?

In contrast to the small number of articles in the preceding frames, 40 articles, or 32% of the sample, addressed cybercrimes committed by adults and against girls. Similar to the preceding frames, however, these crimes are also situated *in* and *as* problematized spaces, with those predators who work with children framed differently than predators who do not. The threat these crimes pose to the social order is reflected in the headlines about sexual predators whose professions bring them in contact with children; their crimes are framed in terms of the contradiction between these crimes and their social positions.

Headlines identify the job titles of cyber-predators whose work brings them in contact with children and teens; however, the role of the Internet in the crime is often delayed until the lead. Some examples of this framing in headlines are "New Charges Against Coach," "Teacher in Pornography Case has County Ties," and "Teacher Arrested on Sex Charge." In the September 28, 1996 *New York Times* follow-up article by Terry Pristin, "New Charges Against Coach," the Internet still is not mentioned in the headline, but rather in the lead: "A soccer and softball coach, arrested last month on charges that he sexually assaulted two young girls, was accused yesterday of molesting four more girls between the ages of 8 and 13 and of transmitting pornographic pictures on the Internet." The Internet also does not appear in the headline of a January 2002 article by Elissa Gootman, but it does appear in the lead: "Teacher Arrested on Sex Charge: A teacher on Long Island has been arrested on charges that he solicited a sexual encounter over the Internet from a 14-year-old former student. The exchange occurred in an online chat room."

In both examples, it is the suspects' positions and alleged crimes that are emphasized in the headlines and not their use of the Internet to gain access to their victims. This distinction is puzzling, since headlines for articles about similar cases against men who do not work with children do not mention position titles, as seen in this example from an August 22, 1997 *New York Times* article "Charge in Internet Sex Case" (Pristin). We learn in the headline that the Internet is somehow involved in the case, but we do not learn until the lead that it involves a minor and a man with no professional connection to children or teens:

> A 39-year-old unemployed computer engineer from Long Branch was charged yesterday by Federal prosecutors with taking sexually explicit photographs of a 14-year-old girl whom he met in an Internet chat room in June. He traveled to Minnesota to bring her to New Jersey but left her at a bus stop when he found out her parents called the cops.

These 2 articles are not only examples of the role of headlines and leads in framing a distinction between online sexual predators who work with kids and those who do not, they also demonstrate the unique nature of reporting these types of crimes in that the primary voices we hear are the official voices of law enforcement authorities and defense attorneys; suspects rarely speak in these stories, and their underage victims will not or cannot. Attorneys were cited as sources in 25 (63%, n = 40) of the articles, while the suspects were cited in 5 (13%, n = 40).

Cyber-Victims: Adults @ risk

Interestingly, articles that were categorized as adults-at-risk featured more youth sources than any of the other frames. The concept of "space" again proved to be an issue here as the location of the computer in the home is made to signify inept parenting; parents supply the tool and the environment for access to their children. Seven articles in the sample (6%, n = 125) were categorized as adults-at-risk, with youths cited as sources in 29% of the articles (2 articles, n = 7). However, experts were the primary sources, cited in 71% of these articles (5 articles, n = 7), with police as the primary sources in 4 (57%, n = 7). Parents were quoted in 3 of the 7 articles (43%).

With language such as "children at play" and "caution" (as opposed to "danger"), the headline of a November 28, 1993 *Washington Post* article, "Caution: children at play on information highway; access to adult networks holds hazards" suggests that maybe Internet crime is not all that serious. The lighthearted tone continues with the lead:

> Genevieve Kazdin, a self-appointed crossing guard on the information highway, remembered the day last September when she found an 8-year-old girl attempting computer conversations with a group of transvestites. Seemingly safe at home, the child was playing with her favorite $2,000 toy, using her computer and modem to make new friends through a service called America Online. (Schwartz, 1993)

The lead presents several themes found in articles warning parents about the dangers of children using the Internet. First, the lead suggests that friendly intervention, in the absence of supervision, can protect children from interacting with adults inappropriately. Second, and more subtle, is the suggestion that children using the Internet unsupervised are only "seemingly safe at home," a suggestion that resonates with parents who are already concerned that their children are not safe *outside* the home.

This theme of parental ignorance continues, even when the interaction is a little more risqué, as seen in this quote from one parent in regard to her thirteen-year-old son: "I would be thinking, how nice that he's spending so much time developing his writing and typing skills, when suddenly he would ask, 'Hey, Mom, what does '69' mean?'" The expert source cited in the article also continues this theme of parental ignorance about the dangers their children face at home: "We call it the IUD syndrome," said Jim Thomas, a University of Illinois sociology professor who studies the computer underground. "Parents are ignorant, technology is ubiquitous, and some of the information is deleterious."

In the *Washington Post* article "Urging Caution in Cyberspace; Va. Presentation Warns Children About Internet Pedophiles" (O'Hanlon, 1998), the headline suggests that children need to be trained to protect themselves from cyber-predators, but the lead suggests otherwise:

> Allyn Worden, 14, of Centreville, finds that Internet chat rooms bring him not only conversations but occasional pornography. "People try to send me pictures every once in a while," he said. "I just delete them." Alyssa Chaffin, 10, of Woodbridge, also uses chat rooms frequently, but so far, she has only witnessed some ugly goings-on, not been the victim of them. "They just call people bad names," she said. "I tell my parents."

Like the previous article, the sources belie the gravity of Internet crimes against children by citing children who have had relatively benign experiences online and have responded to them appropriately—without the help of their parents. Expert sources also send conflicting messages about the role of parents in protecting their children from cyber-predators, as a representative from the National Center of Missing and Exploited Children is quoted: "We're pro-cyberspace," he said. "But for an awful lot of parents, there's a false sense of security . . . Millions of people are coming into our home via cyberspace." As seen in the previous article, it is the parents' ignorance of the extent of the problem that puts their children at risk. But the real frightening information is presented in the very next line as the training program is linked temporally to ongoing raids on cyber-pornography rings:

> Coincidentally, the event took place on the same day that federal authorities announced raids against 200 suspected members of an Internet child pornography ring spanning at least 14 countries. Those arrests are just the latest evidence of an increase in the exploitation of children through the Internet, authorities say.

It is the inability of parents to grapple with the technology, and the disastrous implications of its abuse, that positions adults as victims of the technology—the technology they bring home for their children. The police cited in Tina Kelley's (1998) article "Teaching Parents How to Protect Children on Line" also appear to blame parents' ignorance of the technology for the continuing victimization of their children: "A lot of parents don't understand computers, or how to learn about Web safety when they're told to go to this Web site for more information. We're the generation with the blinking 12:00 on the VCR." Other police sources are more open in their assessment of parental blame for the ongoing threat by pedophiles:

> "Nobody wants to hear it," said Sargeant Russell, commanding officer of the state police computer crimes and electronic evidence unit. "The attitude is, 'It's not my kid going into chat rooms and instant messaging strangers. We're from a well-to-do family.'" Are parents in denial? Some law enforcement officials said yes, they are. For instance, the turnout for parents' discussions about Internet safety at local schools is rarely more than a handful. Most parents don't know their child's screen name or whom they are messaging. (Stowe, 2002)

The November 2, 2003 *New York Times* article by Kate Zernike, "The Watchful Parent, Sentry over Innocence," suggests that parents, even when vigilant, are unprepared to protect their children from cyber-predators: "The morning after the news broke that the headmaster at Trevor Day, a private school in Manhattan, had been arrested on charges of soliciting sex from teenagers online, tearful parents who gathered there expressed not simply shock, but something closer to betrayal." Later in the article, specific language connects this Internet sting case with the Catholic Church sex abuse scandals, further playing on parental fears:

> Trevor Day was hardly an isolated case. The sting that led to Dr. Dexter's arrest resulted in the arrest of other educators, including a popular sixth-grade teacher in Pelham, N.Y. in Westchester County, early this year. . . . The breaking open of the pedophilia scandal in the Roman Catholic Church in the United States over the last two years triggered similar fears about people long seen as figures of respect. (Zernike, 2003)

This connection to loss of trust in individuals occupying respected positions reflects society's already-existing fears—i.e., the cultural resonance component of framing. This connection also provides parents another, tangible, image of sexual abuse, as inferred by the reference to the Catholic Church scandal.

Cyber-Vixens: Police-as-hero

The largest number of articles, 48 out of 125 (38%) were framed as police-as-hero and, not surprisingly, featured police as sources in 73% (35 articles, n = 48). Attorneys were cited in 19 articles (40%, n = 48) and experts in 7 articles (15%, n = 48). The predominance of this frame suggests that cyberspace, like home and school, is gendered and problematic. In headlines, leads and sourcing, these are not stories of "villains" or "victims"—they are law-and-order stories lauding the success of police heroes. Articles featuring headlines such as "100 Subscribers Arrested in Web Child Porn Ring," "FBI Cracks Child Porn Ring Based on Internet; 89 Charged So Far in 'Operation Candyman'," and "20 Charged in Porn Ring after Minors Seen on Web; Some Suspects Used Their Own Children, Authorities Say" demonstrate not only the magnitude of the crimes that police are successfully battling, but also the horror of parental involvement.

Leads for these articles frequently reinforce the police-as-hero images created in the headlines. The lead of the *Washington Post* article "100 Subscribers Arrested in Web Child Porn Ring" elaborates on the details provided in the headline: "One hundred subscribers to child porn web sites have been arrested, with 'many more' arrests expected soon, after a two-year undercover sting operation smashed the largest commercial child pornography ring in the United States, federal officials said yesterday" (Cooper, 2001). This same reinforcement appears in the March 19, 2002 *Washington Post* article "FBI Cracks Child Porn Ring Based on Internet; 89 Charged So Far in 'Operation Candyman'," in which the lead elaborates on the headline by specifying the positions held by suspects: "Eighty-nine people, including two Catholic priests, two police officers, a foster parent and a nurse, have been charged in connection with an Internet-based child pornography ring that authorities have broken up, Justice department officials announced yesterday" (Thompson, 2002).

In the October 7, 2001 *New York Times* article "Setting the Traps to Snare Online Predators" the lead provides more details about the amount of hard work and dedication police have invested in catching cyber-predators:

> After countless hours in the dim glow of computer screens, the police have learned to expect hurdles as they investigate suspected predators who troll the World Wide Web for children. Internet service providers unknowingly discard evidence. Language barriers can delay contact with the police in faraway jurisdictions. And young victims, blurry-eyed with romantic visions, have a hard time believing they were victims at all. (Richter, 2001)

These details, ironically, create eerie parallels between police and the predators they are trying to catch and the girls they are trying to protect. The hours spent sitting before the computer screen, anonymity in discarded evidence, and delays in arrest by police in the distant jurisdictions; this lead could just as easily describe a young victim or a sexual predator. When quoted as sources, however, police and attorneys highlight the magnitude of the problem, justifying continued sting operations: "These are tragedies happening all over the place," said Detective Rory Forrestal, one of Suffolk's first computer crimes investigators. "It's getting worse instead of better." In the same article, Detective James Turner, the newest investigator in Nassau County's three-member computer crimes section, said the scope of the problem floored him as he watched an F.B.I. agent pretend he was a boy during an online training session in Albany (Richter, 2001).

Cyber-Victims: Civil liberties @ risk

Thirteen articles (10%, n = 125) were framed as civil liberties-at-risk. Police and parents were cited as sources in 2 of these (15%, n = 13), experts in 3 (23%, n = 13), and attorneys in 7 (54%). Given the perceived threat to civil liberties evident in this frame, it seems logical that so many articles would feature attorneys as sources. For example, in the June 20, 1999 *Boston Globe* article "Catching Criminals via Net Stirs Debate," attorney Mark Sisti is quoted challenging police efforts: "When it gets to a point where we applaud those individuals in law enforcement who are the best at deception, we should stand back and say, 'Is that what we want from law enforcement?'" ("Catching Criminals," 1999). The question, however, is not allowed to go unanswered as an official source is quoted about the details of the sting: "All I did was go into a (chat) room . . . I had a guy drive from outside Washington, D.C., to Claremont to have sex with a 14-year-old girl," said Claremont police Lieutenant Albuert Stukas, who caught the alleged perpetrator. "In this case, she was fictitious. But what if she isn't?" This civil liberties-at-risk frame presents readers with an image of police and attorneys battling over the use of stings and permits police officials the concluding quote, ending with a question that resonates with our fears about cyber-predators and perhaps convinces us that it may be worth sacrificing our civil liberties.

In another case, a judge went so far as to declare that a "veteran local newsman" could not use the First Amendment in his defense for exchanging child pornography with an undercover cop. The newsman, Larry Matthews, claimed he was researching a story about online child pornography (Castaneda, 1997). When the case went to trial the following year, it was no longer a case about accessing cyber-pornography but rather a case about the scope of the First Amendment. In the April 27, 1998 *Washington Post* article "Writer's Internet Porn Case Tests 1ˢᵗ Amendment," the challenge to our civil liberties is framed as a free press issue championed by an ex-radio journalist, as demonstrated in the lead: "At first, Larry Matthews sat at his Macintosh computer in the office of his Silver Spring home, signed onto America Online and tried to research a story on child smut on the Internet and efforts by the government to police it" (Castaneda, 1998). Unlike other Internet predators, the alleged crime is "research" about "child smut" conducted from an "office" in a home located in a suburb near the nation's capital. In contrast to articles about other cyber-predators, the lead to this article about a journalist accused of exchanging child-pornography over the Internet suggests that his research claim was legitimate, and that our fears about Internet crime are putting our civil liberties at risk:

> I think this is an area of law that has to be watched very carefully. It's always in the name
> of some horrendous evil, like child pornography, that important rights tend to get stripped

away," said Robert Corn-Revere, a Washington lawyer who represents journalistic agencies on First Amendment issues.

In "Jeanine Pirro's Sting: Visiting Chat Rooms to Chase Pedophiles" (Worth, 2001), civil liberties lawyers and activists are cited as likening the sting operations to "witch hunts" that detract attention from "more important crimes" (Worth, 2001). One source, Assistant District Attorney Antonio Castro, was quoted as both an official and a parent: "I am as concerned as any parent out there about protecting our children, but why is the whole focus on this and not on other crimes?"

Conclusion

The elements of cybercrime—anonymous suspects, traumatized juvenile victims, unsuspecting parents, and the invisible nature of the crime—allow for easy news framing. As suggested by Gamson and Modigliani (1989), stories about Internet crimes committed by and against girls resonate with audiences, particularly parents who are unfamiliar with the technology and who may feel guilty about bringing computers (and cyber-predators) into their homes. These stories also enjoy heavy sponsorship by the police, the only sources with access to the details of these crimes and their outcomes. In the case of Internet crime, however, official sponsors also help journalists by providing media-friendly packaging of "sting" operations that can be presented through evocative headlines, reinforcing leads, and official sources.

Unfortunately, these frames, as Iyengar (1991) suggests, may influence the solutions we support to combat Internet crimes against girls. Sacrificing our civil liberties may not appear to be too high a price for anxious parents who lack the power to protect their daughters. As Wright (1986) suggests, these frames may heighten parents' anxiety by highlighting the magnitude of Internet crimes while, at the same time, highlighting the obstacles police face in stemming the tide of Internet crimes against girls. The sources who appear in these frames, however, also serve to further marginalize girls. If sources enjoy status conferral through news exposure (Wright, 1986), the converse is also true; those individuals who do not appear as sources are denied that status. And girls, in particular, cannot afford further marginalization. As McRobbie (1991) explained, girls are already thought to be "invisible" in youth culture (p. 1). Framed as victims who need protection from cyber-predators and from the technology itself, girls are further marginalized by their absence as sources and by the police methods used to protect them. By appropriating girls' identities in cyberspace and, by extension, in the news, law enforcement officials substitute their voices and their experiences for those of girls, effectively making these girls, and the crimes committed against them, invisible to us.

Notes

1. Relevant articles included hard news, soft news, and opinion pieces about Internet crimes committed by and against girls under the age of eighteen. The search yielded 141 articles, however, 16 articles were discarded because they were found to be duplicates of AP wire stories or were about crimes against women over the age of eighteen who were referred to as "girls" in the article.

2. Data from this study comes from a project funded by the Mary S. and Christian L. Lindback Foundation exploring juvenile crime news frames in *The Philadelphia Inquirer, The New York Times, The Los Angeles Times, The Boston Globe,* and *The Washington Post* from 1990–2002. Articles were downloaded from Lexis-Nexis using the search terms *Internet, computer, crime,* and *w/5 girl* to find all articles about computer-based or Internet crimes with the word *girl* within five words of crime in these newspapers during this timeframe.

3. Articles in the *girls-as-risk* frame focused on Internet crimes committed by girls, providing details of the crime and information about the girls themselves. Articles in the *girls-at-risk* frame focused on the victims and their responses to the crimes committed against them through the Internet. Articles in the *adults-as-risk* frame focused on the adults who use the Internet to commit crimes against girls. Articles in the *adults-at-risk* frame focused on the technological naiveté of parents and their fears about Internet crimes against children. Articles in the *police-as-hero* frame focused on the police actions, training, frustration and/or triumphs in resolving Internet crimes. Articles in the *civil liberties-at-risk* frame addressed questions about police stings as infringement on the civil liberties of suspects. Articles in the *other* frame did not directly address Internet crime against girls. Instead, these articles presented speculation about the contribution of media and Internet use to non-Internet crimes.

References

Allan, S. (1999). *News culture*. Philadelphia: Open University Press.

Barthes, R. (1982). Myth today. In S. Sontag (Ed.), *A Barthes reader* (pp. 93–149). New York: Hill and Wang.

Bird, S. E., & Dardenne, R. W. (1988). Myth, chronicle, and story: Exploring the narrative qualities of news. In J. W. Carey (Ed.), *Media, myths and narratives: Television and the press* (pp. 67–86). Newbury Park, CA: Sage.

Castaneda, R. (1997, September 4). Radio reporter charged in child pornography case; newsman says he was researching a story. *Washington Post.* Retrieved December 20, 2002, from Lexis-Nexis.

Castaneda, R. (1998, April 27). Writer's Internet porn case tests 1st Amendment. *The Washington Post.* Retrieved December 20, 2002, from Lexis-Nexis.

Catching criminals via Net stirs debate. (1999, June 20). *Boston Globe.* Retrieved December 20, 2002, from Lexis-Nexis.

Children Now (2001). *Local TV news distorts real picture of children, study finds.* Retrieved March 2, 2004, from http://www.childrennow.org/newsroom/news-01/pr-10–23–01.cfm

Cohen, B. (1963). *The press and foreign policy*. Princeton: Princeton University Press.

Cooper, G. (2001, August 9). 100 subscribers arrested in Web child porn ring. *Washington Post.* Retrieved December 20, 2002, from Lexis-Nexis.

Dorfman, L., & Woodruff, K. (1998). The roles of speakers in local television news stories on youth and violence. *Journal of Popular Film and Television, 26*(2), 80–86.

Ericson, R. V., Baranek, P. M., & Chan, J. B. L. (1991). *Representing order: Crime, law, and justice in the news media.* Toronto: University of Toronto Press.

Finkelhor, D., Mitchell, K., & Wolak, J. (2000). *Online victimization: A report on the nation's youth.* Arlington, VA: The National Center for Missing and Exploited Children.

Gamson, W. A., & Modigliani, A. (1989). Media discourse and public opinion on nuclear power: A constructionist approach. *The American Journal of Sociology, 95*(1), 1–37.

Gans, H. R. (1979). *Deciding what's news: A study of* CBS Evening News, NBC Nightly News, Newsweek, *and* Time. New York: Vintage Books.

Girl, 15, arrested making drugs at home with Internet recipe (1997, October 21). *Los Angeles Times.* Retrieved December 20, 2002, from Lexis-Nexis.

Gootman, E. (2002, January 14). Teacher arrested on sex charge. *The New York Times.* Retrieved December 20, 2002, from Lexis-Nexis.

Gootman, E. (2002, November 26). On stand, girl recalls week of captivity and rape on L.I. *New York Times.* Retrieved December 20, 2002, from Lexis-Nexis.

Hayes, B. (1997, October 22). California and the West; Internet drug recipe raises officials' fears. *Los Angeles Times.* Retrieved December 20, 2002, from Lexis-Nexis.

Iyengar, S. (1991). *Is anyone responsible? How television frames political issues.* Chicago: The University of Chicago Press.

Kelley, T. (1998, August 20). Teaching parents how to protect children on line. *New York Times.* Retrieved December 20, 2002, from Lexis-Nexis.

Kilgannon, C. (2002, May 22). Slain girl used Internet to seek sex, police say. *New York Times.* Retrieved December 20, 2002, from Lexis-Nexis.

Koch, T. (1990). *The news as myth: Fact and context in journalism.* Westport, CT: Greenwood Press.

Lipari, L. (1996). Journalistic authority: Textual strategies of legitimation. *Journalism and Mass Communication Quarterly, 73*(4), 831–857.

Mackey-Kallis, S., & Hahn, D. (1994). Who's to blame for America's drug problem?: The search for scapegoats in the "war on drugs." *Communication Quarterly, 42*(1), 1–20.

Males, M. (1994, March/April). Bashing youth: Media myths about teenagers. *Extra!* Retrieved March 3, 2004, from http://www.fair.org/extra/9403/bashing-youth.html

Mazzarella, S. R., & Pecora, N. (2002, November). *Girls in crisis: Newspaper coverage of adolescent girls.* Paper presented at the meeting of the National Communication Association, New Orleans, LA.

McManus, J., & Dorfman, L. (2002). Youth violence stories focus on events, not causes. *Newspaper Research Journal, 23*(4), 6–20.

McRobbie, A. (1991). *Feminism and youth culture: From* Jackie *to* Just Seventeen. Boston: Unwin Hyman.

Neuman, W. R., Just, M. R., & Crigler, A. N. (1992). *Common knowledge: News and the construction of political meaning.* Chicago: University of Chicago Press.

O'Hanlon, A. (1998, September 3). Urging caution in cyberspace; VA. presentation warns children about Internet pedophiles. *Washington Post.* Retrieved December 20, 2002, from Lexis-Nexis.

Pristin, T. (1997, August 22). Charge in Internet sex case. *New York Times.* Retrieved December 20, 2002, from Lexis-Nexis.

Pristin, T. (1996, September 28). New charges against coach. *New York Times.* Retrieved December 20, 2002, from Lexis-Nexis.

Richter, A. (2001, October 7). Setting the traps to snare online predators. *New York Times.* Retrieved December 20, 2002, from Lexis-Nexis.

Roug, L., & Harris, B. (2000, April 8). Teen who had bomb material is arrested; Irvine Girl, 16, is in custody after discovery of materials and instructions triggered evacuation of apartment complex.

The youth tells police the chemicals were for use in a film project. *Los Angeles Times*. Retrieved December 20, 2002, from Lexis-Nexis.

Schwartz, J. (1993, November 28). Caution: Children at play on information highway; access to adult networks holds hazards. *Washington Post*. Retrieved December 20, 2002, from Lexis-Nexis.

Shepard, A. C. (1994). Identifying juvenile suspects. *American Journalism Review*, *16*(5), p. 14.

Shepard, A. C. (2002). Young lives, big stories: Crime, abuse and little context dominate coverage of children's lives. *American Journalism Review*, 24(2), pp. 1–5.

Stevens, J. E. (2001). *Reporting on violence: New ideas for television, print and Web*. Berkeley, CA: Berkeley Media Studies Group.

Stowe, S. (2002, May 26). Police ask parents to hover when their children log on. *New York Times*. Retrieved December 20, 2002, from Lexis-Nexis.

Thompson, C. W. (2002, March 19). FBI cracks child porn ring based on Internet; 89 charged so far in 'Operation Candyman.' *Washington Post*. Retrieved December 20, 2002, from Lexis-Nexis.

Tuchman, G. (1978). *Making news: A study in the construction of reality*. New York: Free Press.

White, J. (2001, February 6). Police say girl sent AOL threats to self; teen allegedly tried to blame others. *Washington Post*. Retrieved December 20, 2002, from Lexis-Nexis.

Worth, R. (2001, July 15). Jeanine Pirro's sting: Visiting chat rooms to chase pedophiles. *New York Times*. Retrieved December 20, 2002, from Lexis-Nexis.

Wright, C. (1986). *Mass communication: A sociological perspective* (3rd ed.). New York: Random House.

Zernike, K. (2003, November 2). The watchful parent, sentry over innocence. *New York Times*. Retrieved December 20, 2002, from Lexis-Nexis.

Zoch, L. M., & Turk, J. V. (1998). Women making news: Gender as a variable in source selection and use. *Journalism and Mass Communication Quarterly*, *75*(4), 762–775.

2

What are gURLs Talking about?

Adolescent Girls' Construction of Sexual Identity on gURL.com

Ashley D. Grisso and David Weiss[1]

Introduction

How do teenage girls feel about themselves as sexual beings? How do they talk about sex? How does their discourse around issues of sexual acts and sexuality contribute to the construction of their individual sexual identities? And how might an Internet community work to inform such identity development? This chapter explores these issues through an in-depth look at the discourses around sex taking place on gURL.com, the largest and most popular Web site targeting teen girls.

As McRobbie (1991), Pecora and Mazzarella (1999), and Wyatt (2000) have pointed out, until the last two decades of the twentieth century, the vast majority of the research purporting to study "adolescence" actually focused almost exclusively on adolescent *boys* and their development. However, girls negotiate their own gender-specific transitions. In order to investigate the issues of greatest importance to girls in this transitional age group, it is important to hear them speak using their own words (Edut, 1998; Gilligan, 1982; Mazzarella & Pecora, 1999; Shandler, 1999). Communicating in their own words helps girls develop not only their sense of self and identity but also allows them to construct their own social reality as members of peer groups (Berger & Luckmann, 1967; Blumer, 1969; Kroger, 1996; Mead, 1967). Such identity development—including, importantly, gender identity and sexual identity development—can be seen as *performative*: in the very act of talking or writing about her sense of (gendered/sexual) self, a girl does not merely describe her identity, but actually brings it into being (Austin, 1962; Butler, 1990, 1993).

Along with other researchers, we believe that girls will be most free to explore and construct their identities and express feelings about the issues of greatest importance to them when they are in a space they consider safe—that is, free from the potentially judgmental or inhibiting influence of adults or male peers. As Bentley (1999, pp. 219–220) observed,

> Girls need safe spaces . . . outside of the traditional spaces where girls interact with the larger culture . . . They need spaces where they can know what they know and try new identities without self-censoring. Without safe spaces, girls will not be fully able to discover who they are and who they would like to become.

While Bentley referred primarily to literal, physical spaces, the existence of safe spaces online is just as valuable to contemporary girls. Girls' Web sites—that is, sites designed expressly for and about girls, in which their active participation and candid attitudes, fears, hopes and concerns are not only featured, but encouraged—are especially rich and revealing sources of such data. Yet, as Stern pointed out in one of her publications about girls' personal home pages, "only a handful of researchers have paid attention to existing and potential venues for self-expression. Thus, we still know relatively little . . . about what girls at the edge of adulthood have to say about their lives, whom they want to listen to them, and how they regard their expression" (2002, p. 224).

Stern's research on girls' online self-disclosure (1999, 2002) has broken important new ground. As she has argued (2002, pp. 228–229),

> Teens have the potential to interact online with more people like themselves . . . They may find people online who can accept parts of them that their other acquaintances cannot, or they can discuss topics that other friends and family avoid . . . address taboo or unsavory topics, and . . . experiment with different self-presentation styles . . . [and] facilitate girls' self-disclosure.

We believe that the girl-authored sections of the gURL.com Web site offer girls many of these same uses, gratifications, and benefits.

Much of what we will be looking at in this chapter reveals girls' construction of their identities as sexual beings. As Heilman noted, "each girl is responding to the need to have an identity, a self, in a highly demanding environment . . . Identity is much more complex than the simple labels we use that refer to life stage, class, and gender" (1998, p. 184). The development of a healthy *sexual* identity is particularly important for teens, but this truth is often ignored. As Holloway and LeCompte observed, "the curricular evasions of and silences about the body, feelings, and sexual politics reinforce traditional gender roles and rarely provide support to help girls make decisions about complex personal and moral dilemmas in their lives" (2001, p. 391).

Indeed, as Irvine noted, "in our society, adolescent sexuality is a social problem" (1994, p. 5). But adolescents, as Sutton, Brown, Wilson, and Klein (2002) pointed out, are "hungry for sexual information. This is a time in their lives when the formation of a healthy sexual identity is vital to their mental health . . . Identifying sources from which American adolescents are most likely to acquire such information is crucial" (p. 49). A Web site such as gURL.com is an important source of information, providing as it does an antidote to what Holloway and LeCompte (2001, p. 390) critiqued as the "curricular evasions and silences" so tragically prevalent in public discourse about adolescent girls.

About gURL.com

gURL.com is a Web site targeting girls aged thirteen and above. Originally developed in 1996 by Esther Drill, Heather McDonald, and Rebecca Odes as a project for New York University's Interactive Telecommunications Program, the site has been bought and sold several times during its seven-year life. In August 2003, the Manhattan-based women's media company iVillage, Inc. acquired the site from Primedia, Inc., publisher of teen magazines *Tiger Beat, Teen Beat,* and *Seventeen.* Primedia had owned gURL.com for only two years, having purchased it in 2001 from teen-targeted mail-order clothing retailer dELia*s Corp., who had purchased the site from its founders in 1998 (Jain, 2003; Joyce, 2001; Morrissey, 2003). Despite gURL's frequent changes of ownership, co-founders Drill and McDonald are still at the helm, having moved to iVillage after the 2003 acquisition.

The gURL.com site is the largest, most diversified of the numerous adolescent-girl-targeted sites currently accessible on the World Wide Web. It is also the site that has received and continues to receive the most attention in the popular media, not only for its success, its content, its outlook, and its resultant wide appeal to girls but also for its well-received (Ivinski, 2000; Lodge, 2000) companion book, *Deal with It!* (Drill, McDonald, & Odes, 1999), written by the site's creators. As *Publishers Weekly* noted, gURL.com and *Deal with It!* take "a holistic approach to . . . perennial teen concerns: changing bodies, emotions, desires, and lives . . . in a frank, nonjudgmental tone" (Zaleski & Gediman, 1999, p. 72).

The gURL.com site addresses complex and important subjects head on rather than taking the tentative approach characteristic of most other sources of influence in adolescent girls' lives. As stated on the site's "What is gURL?" page (gURL.com, 2003), gURL is a different approach to the experience of being a teenage girl. We are committed to discussing issues that affect the lives of girls age 13 and up in a nonjudgmental, personal way. [W]e try to give girls a new way of looking at subjects crucial to their lives . . . Our content deals frankly with sexuality, emotions, body image, etc. If this is a problem for you, you might not like it here.

While many parts of gURL.com are authored by adult contributors, in this chapter we are concerned only with sections of the site written by its teen girl members. Specifically, we will be analyzing discourses around issues of sex, sexuality, and sexual identity appearing in girls' posts on either of two of the site's bulletin boards: "Shoutouts: Sex" and "Dig or Dis: Premarital Sex."[2]

A Peek into the Pow(d)er Room: gURL.com as Haven

The paradox of the Internet as a site of communication is its ability to atomize individuals even as it brings them together (Weiss & Grisso, 2003). The role of the Internet as a builder of community (Mantovani, 1996) can be seen on gURL.com as well. Those girls who choose to participate in discussions about sexual behavior, and make themselves vulnerable by doing so, may discover that they have found a safe place to do so:[3]

NEED_HELP_LOL: *i no this is sad but what is swallowing???*
AngelicDevil88: *it's when a boy cums in your mouth and you swallow his semen*
NEED_HELP_LOL: *thnx i feel like an idiot asking that question*
AngelicDevil88: *better that u ask it here than not know – – – that's what we're all here for! gg me*[4]
. . . i won't laugh at you. knowledge is important.

Indeed, it is apparent that one of the implicit norms of gURL.com is that one's fellow gURLs are on the site to help and support each other. When a member ignores or defies this norm, she may find herself chastised for doing so:

eyedream: *Excuse me, but glowbug had a valid question and you had no right to insult her like that. This is a message board for all girls. Girls have it hard enough already, don't they? let's not make it worse by picking on each other.*

The comments offered by eyedream and AngelicDevil88 illustrate what we have come to think of as the "pow(d)er room": a special space reserved exclusively for girls, as the school powder room has always been, and at the same time, a place where girls find not only safety but (girl) power, a site in which they can build, discover, and assert identity and agency. By using this term, we are reclaiming/empowering what some might consider a derisive or dismissive term. Given gURL.com's status as "pow(d)er room," we were not surprised to find a girl with the screen name *ilovegurldotcom*.

"What the Heck Is Premarital Sex?"

Indeed, quite a number of gURL.com members use the site to ask and learn about the most basic sexual facts. Many of the questions asked by girls who, we assume, have had little to no sexual experience of their own, concern *definitions*; girls hear sexual terms but don't yet know what they denote:

> jessi112191: *what is a blo job? i would really like to know so that i will know what they are talking about? HELP ME !!!!!!!!!!*
> mrsmathersllp: *i know what doggy style is but what is missionary?*
> ImSexyInPink: *can somebody please help me. whats 4-play?*
> Sk8eR_cHicK121: *wut r the bases?*
> M-O-LEE: *WHAT THE HECK IS PREMARITAL SEX!!!!?????? whateva it is it doesn't sound good*

Other girls ask "how to" questions:

> cheerinQT33: *how exactly do you give an erotic massage so it's different from a normal massage?*

Still others ask questions or give (bad) advice about topics which ideally would be covered in a sex-education class:

> BBallhunni02: *how do you have sex?*
> RubMyPoPtarT: *Yes you can get pregnant if he cums near your vagina, or if it even touches your pubes. Sperm can live up to 7 days. Get tested.*
> Prettyinpink200: *So u want to know what a blow job is. Its basically putting your mouth on his dick and sucking and blowing.*

Other girls reflect a startling combination of sophistication and naïveté; they discuss "adult" topics but approach them from a place of innocence:

> Princess41589: *when we do it for the first time do we just lay there with our legs open? or do we have to do something? like does the guy know which whole to put it in if we don't, even if it's his first time as well . . . i'm confused.*
> Lemonlover11: *is it ok to use a back vibrator on your vaginal area?*

Some girls even offer general advice-giving services:

> surfergirl1733: *hey gurls!!! if you ever need advice about something or if you have questions or whatever gimme a shout on [email address] or gg me. I've gone threw a lot of stuff with guys, emotions and well pretty much everything. so yeah, feel free to add me. ChOw!!!*

Cumming and Becoming

gURL.com is not just a sex-advice column for its members. The sex discussion board serves as a site for girls to explore and to begin to define themselves as sexual beings. Much of the "becoming" process played out on the site concerns girls' engaging with virginity, specifically, the mixed feelings they experience as they find themselves torn between the desire (in some cases) to remain virgins while feeling pressure to become sexually active. We would argue that, regardless of a girl's "status" (virgin or otherwise), the mere act of exploring the possibility of entering the world of the sexually active is the beginning of the construction or performativity of sexuality.

Some girls are openly ambivalent about their feelings:

> MonkeyButt145: *ok i want to have sex really bad like u don't even know and i just want to be a virgin too*
> goodcharlotte26: *i don't want to regret having sex but sometimes i wonder if i should . . . its really confusing*

Yet there are also quite a few gURL.com members who are proudly, even defiantly, virginal and intend to stay that way until they are married. Some, but by no means all of these girls offer religious or moral justifications for their pro-virginity position:

> Niliwen: *For one thing, it's VERY IMMORAL!! (I'm Catholic, and i believe that sex should be saved for marriage). Besides, you will be forming unnecessary emotional baggage w/ the people you slept with*
> zionsdaughter: *Sex is serious and God didn't make it so you could screw around with some dork just trying to get some!!! and what does having sex before marriage make you? a SLUT, HOE, and one big SKANK!!!!*
> Girl4God28: *Well, i say it's wrong. The bible says that sex is for two people who are married. i mean, if you wanna upset god, GO RIGHT AHEAD.*
> authority_sux: *im 13, and i don't think its right. you have so much ahead, don't ruin it by getting dumped by sleeping wit sum 1 or becoming the skool class slut. cause the word will get out!*
> dmscheerchik: *i think only skanks, bitches, and hoes have sex b4 theyre married.*

We were struck by the persistence of the age-old double standard: sexually active girls are "sluts, hoes, and skanks," while sexually active boys are not stigmatized. Related to this is the fact that girls use language that indicates that premarital sex for girls involves giving something away or losing something, while for boys the same action is seen as getting some(thing).

Many of the girls on the site seem to have made the decision that they will soon become sexually active, a decision they are not taking lightly. Many are scared and worried. They are also highly conscious of their lack of experience, seeing it as a source of embarrassment:

BabyGurl04082: *i need some insight on sex since im 13 and ill be doing it soon*
Beachbabe_89: *Hey gurls. OK i'm only 13 years old and i'm like desperate to have sex. Am i too young?????*
dani8811: *i've never been fingered or had sex b 4 and i don't even use tampons so how will i be able to have sex!? because i really wanna start, but i'm afraid it will hurt a lot and the guys penis won't fit . . . that would be really embarrassing . . .*

Sadly, many girls don't have a positive sense of themselves, and their lack of confidence is compounded by fears of boyfriends' judgments:

mizz_ApPlE: *my man wants to go down on me and he want to do it doggy style. wat exactly is that? i wanna know how to prepare for these things like cuz i don't wanna smell down there when he go down on me and plus isn't doggy style in da but and would his penis come out wit feces on it . . . LOL*

Girls' lack of self-confidence around these issues is tied to their worries about how boys might perceive them. This can be seen in the expressions these girls use for body parts and sexual acts, expressions more traditionally associated with "men's language" (Johnson, 1997; Lakoff, 1975) including the language of male-originated and male-targeted rap songs and pornographic movies.

Despite girls' feelings of uncertainly and inadequacy, the community of gURL.com can offer reassurance, as can be seen in the responses to coolchica99's posting:

coolchica99: *my boyfriend* [of 5 months] *says he wants us to go furthur (oral) and i think i'm ok with this, i'm just paranoid about my body. He's experienced, so knows what girls are 'meant' to look like. What if i'm wrong? Do you think we could do it in the dark or something?*
sarahmarie:[5] *you do not look wrong. every girl's body is different. guys are just happy to get to see your body. a lot of girls here seem really self-concious about their bodies. *don't be*. any decent partner will love your body and moreover, respect you. be confident. leave the lights on, it's much more fun to see each other's bodies. don't have sex if you don't feel comfortable with your own body - get comfortable with it yourself, first*

"U R a LESBO"

Learning about sex acts—particularly those that one performs or wishes to perform oneself—is one of the first steps toward constructing one's sexual identity (Christopher, 2001; Irvine, 1994). As we saw with coolchica99 (above), girls use gURL.com's message boards as a place to discover facets of their own sexual identities. While most of these identities are heterosexual, this is not the case for all gURL.com members. Girls who are beginning to feel same-sex attractions do, on occasion, share their questions and concerns on the site. Some of these girls distance themselves from the objects of their curiosity; others are further along the path to self-knowledge, self-acceptance, and identity around such attractions:

BlondeBabe1313: *this is gonna sound stupid, but how do lesbians or gay people have sex??? im not lesbian or gay but just wanted to know*

Crystal_Keeper: *i think i want to be in bed with another girl and make love to her. what should i do? i don't know if i'm a lesbian or not. i'm attracted to boys and girls. if i see a pretty girl, i want them to touch me and the same about boys.*

TakeMe2: *when me an my neighbor were little. we would go under the covers naked, and hold eachother, and kiss, the full monty. even fiddling with clits, etc. and we masterbated together, it was soooooo much fun, i'm pretty curious*

There is often quite a range of responses to such questions. These responses reflect not only the values of the responders but in some cases they construct sexual identity as well. Gangstette's response to TakeMe2's post (above) serves to construct Gangstette as *other than* a lesbian; in other words, as heterosexual. The responses from other girls are not only more tolerant but also indicative and constitutive of their own sexual identities:

Gangstette: *i know exactly what this case is . . . U R A LESBO!!*

CoLdMnKy: *r u sure your attracted to her in that way? maybe u just really like hanging out with her. even if you're suddenly bi just go w/ it. it's the best of both worlds. n its ok not gross. just normal. i know a lot of bi girls and guys.*

Soccerstar291: *hey its OK u might get over it but i dunno cuz i'm the same way.*

"OK so im having sex with this guy . . ."

While most of the girls whose discourse has been discussed thus far indicate in some way that they are sexually naïve, many of their fellow gURL.com members construct their online identities in such a way as to suggest greater sexual precocity. Whether these teens have actually had all the sexual experiences they claim to have had may not matter; whether factual, fictional, or somewhere in between, their narratives capture the performativity of their sexual identities; indeed, as Turkle (1995) has pointed out, the Internet is frequently a forum for the creation, and even fabrication, of identities. A number of girls construct and share identities characterized by confidence and comfort around their sexual practices; others, who may be equally experienced, are ambivalent about whether the fact that they are actively sexual benefits them. This excerpt from a post from Luv_slave_13 is representative of the ambivalence expressed by many of the girls:

Luv_slave_13: *whenever i do something sexual i feel guilty about it afterwards. i haven't had a lot of sexual partners, only people i really care about, so you'd think it would be a good experience. i feel like i'm ready for sex too, like, i feel good when i'm doing it, i feel well educated about birth control, i'm not nervous bout it, but i feel guilty after, like ive done something bad and i know sex isn't bad even if you're a teenager, 'cause it's human nature. help.*

Other girls are more seriously troubled by their sexual pasts:

> Candycanelady: *i just had sex for the first time about 3 days ago. Every time i think about it, my stomach feels weird and i feel shaky and cold*
>
> rockerby42: *you might think that giving head is just so glamorous, but its just not. i lost my gag reflex but in place of it i cry when im giving oral to my fiancé.*
>
> hennessy26: *oh my god, i have sex constatly. this started 5 and a half years ago. it's always with different guys. i've not done one guy twice. no one seems to notice, except the guys i have sex with. one time, i went through a 36-pack of condoms in 4 weeks. i'm 22 now, so it started when i was 17. i'm like, a professional. Is this getting serious? should i stop. HOW do i stop, LOL.*

Some girls who portray themselves as sexually active express concerns about lacking certain technical expertise, particularly regarding acts that will please their partners:

> kinkykisserXOXOXO: *can you gurls give me some tips on how i can really turn him on if i give him head. i'm nervous that i'm not gonna do something right.*
>
> button_kisses: *Ok so im having sex with this guy and the sex is great but he wants me to go on top. i want to do that but i don't know how. how do you move and how far up and down do you go? please help me. i can't keep avoiding it forever.*
>
> angelgurl_13: *this is kinda embarrassing but i need to ask . . . i totally freak out every time i touch him . . . ya no . . . down there, intimately . . . i feel like SUCH an idiot. i know it doesn't matter to him but it does to ME . . . i want to pleasure him, too. i mean sex wouldn't freak me out, but we're waiting on that, its bj's and hj's and messing with that area of his body skin-to-skin freaks me out.*

One girl, who seems completely at peace with having sex, is less comfortable about the facts of her own female sexuality:

> HugzNKissez24: *OK . . . so i had sex w/my bf last week . . . it hurt a bit but it wuz amazing!! anyway . . he says he wants to do it this weekend . . my period came today . . . well . . i don't want to have my period and have sex of course!! but i don't wanna tell him that i can't have sex just cuz my period*

Several girls responded to HugzNKissez24, reassuring her that it was perfectly okay to let her boyfriend know about her period, that she had nothing to be ashamed of, and that they supported whatever decision she decided to make. Such community building around a girl's expressed concerns is, in fact, common on gURL.com.

It appears that some girls feel a great deal of pressure to present themselves as more sexually experienced than they actually are, or at least to test the waters, so to speak. One girl asked a rather naïve question which led us to believe that she was fairly inexperienced:

> Aliecat25: *i'm still technically a virgin so i was just wondering if there was any chance of getting pregnant through anal sex?*

Several days later, however, this same girl posted a long, explicit narrative verging on a pornographic fantasy in which she was the star.

Aliecat25: me & my 2 best friends snuck out 2 my friend's neighbor's house cause he was having a party. we were cold so we went in the hot tub in our bras & thongs. well one of my brothers friends Andy got in (he was really stoned & he was NAKED :-) yeah well he was "cold" so he wanted us all 2 cum sit on his lap so i was on his right leg Desi was on his left leg & Sam was right in the middle!! every 1 else went inside so Andy started 2 rub my leg & stuff, i stayed on his lap and he started fingering me! :-) WOW. it was gr8! he was fingering me 4 almost an hour then me & Andy started making out & WOW he's a gr8 kisser! he has a tongue ring & he knows exactly what to do with his tongue! i guess i like him & all but he has a gf he, yeah my bad! plus he's a player oh well that Monday @ school he started telling all his friends what i let him do to me which i didn't mind until one of them told my brother! . . .

Aliecat25's story, to this point, seemed plausible. However, as it continued, it appeared that she chose to construct herself as a sexually precocious young woman. Was this in response to her new reputation? The narrative continued:

Aliecat25: me & Desi are probably going over 2 her neighbors house again 2morrow night. So im probably gonna get stoned & drunk & give Andy & a bunch of other guys hand jobs or blow jobs so it should b FUN stuff unless my brother shows up! yeah well i just wanted 2 tell that story cuz i had sooooo much fun. almost 4got Andy was trying 2 get me up the ass but it wasn't really work-ing (he was 2 stoned). he he maybe 2morrow I'll get there b4 he's stoned out of his mind! ha ha. if u wanna give me suntin fun 2 do w/sum of the guyz @ the party i'm waiting 2 hear from ya!!! love yall!!!!!!!

This entire story, particularly the second half, illustrates Aliecat25's performance of sexual identity. Whether she actually did all the things in the hot tub that she claimed to have done, at some point it appears that she decided to at least play the part for her fellow gURLs online. Perhaps more telling is the tension between Aliecat25's self-portrayal as sexually eager and the non-agentive language she uses: Her sexual activity with Andy becomes not something that she (or they) did, but rather is "what i let him do to me."

Still, a number of the girls on the site took a dim view of Aliecat25's actions, and were quick to share their concerns and judgments with her.

Pretty_in_punk3: can we say SLUT!
LsBaByGiRl: Hun, not to be rude it sounds like u had a great time but ur makin urself kinda sound like a slut so watch what cha say i wouldn't want u2 end up with a bad reputation on here.

Aliecat25's response to the advice proffered by LsBaByGiRl provides further evidence of the constructedness of her identity as a sexual adventurer:

Aliecat25: Thanx 4 the advice. i know people may think i'm a slut! what the heck maybe i am but whatever. i'm not really that slutty. that's pretty much the most i've done w/guyz. but thanx.
LsBaByGiRl: Sry i didn't mean 2 sound like i was callin u a slut i jus don't want other ppl 2 call u 1 ya kno?

We found this dialogue revealing in its reflection of girls' attitudes toward a peer who would so openly and explicitly recount a sexual adventure (whether fictional or not) and express the intention to repeat such actions. Despite the fact that gURL.com can often be a non-judgmental, safe space, Aliecat25, with her bravado, seems to have crossed a line. But what, specifically, bothered the other girls? Alie's agency? Her story's explicitness? Her self-construction as a slut, a word that she did not use herself, but which her (reported or intended) actions performed for her? In any case, the other girls' responses reified traditional patriarchal norms of sexual propriety: Even on gURL.com, if a girl constructs an identity that is *too* avidly sexual, her peers will rein her in. This was confirmed in another dialogue:

> SoccerLuva931:*Hi. i'm 12 and I've done it 43 times. Is this normal?*
> WyteChocolate1: *gurl, that is not NORMAL . . . u need to watch out. i know it's very hard but you need to slow down . . juss slow down because u r going 2 catch sumptin . . . if you really need pleasure dat bad u needa use a dildo*

Interestingly, not all such elaborate tales of sexual escapades generate censure from other members. Greek_Goddess89, a girl who posted a similarly lengthy, pornography-inspired story, received virtually no response, judgmental or otherwise, following her posting, which opened with a long set-up about her parents leaving her alone at home and then continued as follows:

> Greek_Goddess89: [. . .] *i wanted it sooo bad so then i remembered my hot neighbor. from his room we can see each other perfectly so i called him and told him to look at me through his window. i ran up to my room and saw him looking so i slowly started taking off all my clothes and started amsturbating in front of him. well within seconds he was at my door so we went upstairs and started taking his clothes off and then we went into my hot tub and he sat down and he told me to sit down on his lap. i placed myself in the middle and we start going at it really hard. we started humping like crazy. by the way those little massagers in the hot tub are EXCELLANT when doing wut we were doing. we traded positions. then he got down and started giving me head. it was wonderful. then we started thrusting hard on the floor [. . .]*

Greek_Goddess89's story seems to be lifted from the pages of a pornographic novel, both in terms of its narrative elements and its language. Basing one's own personal narratives on "scripts" provided by media texts is not unusual (Christopher, 2001; Silverman-Watkins, 1983); indeed, as social learning theorists (Huesmann, 1986; Perse, 2001) have shown, scripts, "schemas about courses of action that are stored in memory, when retrieved . . . become guides for behavior" (Perse, 2001, p. 205). This phenomenon, in fact, was even commented upon by another gURL.com member, ren_oblos, who observed *"they hear in music and television guys talking bout tight pussies . . . and then they just kinda copy the behaviour."* Bringing mediated sexual content into "real life" has long been the province of boys and men; what we found striking was seeing girls appropriate sexual "scripts" in this way, adopting not only what has traditionally been considered "male language" but also male ways of constructing identity.

While Greek_Goddess89 claims agency—in her story, she calls her neighbor and initiates all contact—she recounts her narrative mechanically. While her intention may very well have been to construct an identity for herself that could be categorized as sexually experienced, and while her story does result in her positioning herself as a subject, the artificiality of her tale appears to have alienated almost everyone else on the message board. The one response posted to Greek_Goddess89 simply asked "what was it like?" Greek_Goddess89 did not reply, perhaps unable to actually answer this question.

At least one gURL.com girl seems to be appropriating another traditionally male behavior: having a "fuck buddy," that is, a sexual partner who may not be a love interest:

> Brea_Babe: *i have a sorta fukc friend . . . and i made the mistake of asking what we would do if one of us ended up liking the other*

Where are teen girls today getting the message that it is acceptable (or, at least, non-problematic) to have sex without romantic attachment? We suspect that the media are sources of such "scripts," both implicit and explicit.

If It Feels Good, (Should I) Do It(?)

One of the most troubling features of girls' discourses around sex, sexuality, and sexual identity that we observed is the minimal emphasis girls place on their own pleasure. Many gURLs, particularly those seeking technical advice, seem consumed with pleasing their male sexual partners, usually at the expense of their own enjoyment:

> Lily_Frank: *When my bf and i r having sex and i'm on top i don't think he enjoys it as much. Wat can i do to make him like it more?*
> BrokenUpInside: *i no what positions feel best for me but not which ones feel best for my b/f. do u guys know which ones provide the best friction, tighness, etc . . . ?*
> lovemetender13: *can u guys give me a few tips on how to give my bf a hard-on.*
> kinkykisserXOXOXO: *can you gurls give me some tips on how i can really turn him on if i give him head. i'm nervous that i'm not gonna do something right.*
> lilbaby143012: *my b/f and i are having sex, he loves it . . . and cums really easy but i can't seem to get wet*

The girls quoted above fail to express any desire of their own and, more important, fail to convey any sense of personal agency. By contrast, AntiCheesePunk (below) is caught between wanting intimacy with her boyfriend and the growing fear that his interest in her is strictly sexual. Even more troubling is the feeling of worthlessness that this situation evokes for her.

AntiCheesePunk: *tomorrow is me and my boyfriend's 5 month anniversary. i'm thinking of watching movies, cuddling, and laughing. He's thinking of doing more sexual things. i DON'T want to do stuff. i just want to cuddle and have a good time. We ended up getting in a big argument over this. Please help me. It kinda hurts to think that maybe that's all he wants from me. i hate feeling worthless*

Even PLAYboyzRmyTOYZ, who openly expresses sexual desire and agency, still has reservations about having intercourse with her new boyfriend.

PLAYboyzRmyTOYZ: *my new bf is comeing over he is really hot and has a really nice dick! i sucked him a few times already and played with him a few times. hes sick of being a virgin and he wants to screw me when he gets here. im really horny so i wouldn't mind but i don't normaly sleep with the guy after a week of going out with him. he just is so hot and has this great body and way of hipnotizing me with it! i will give him a blow job and hand job but i don't know if i should screw him yet*

Even more interesting, and poignant, is PLAYboyz's lack of concern for her own pleasure and the fact that she discusses the various sexual favors she plans to "give him" in fairly pornographic terms. The subject position she constructs for herself appears to combine the traditional female romantic role (focusing on her partner's pleasure) and the traditional male sexual role (sexual avidity and "horniness").

What's Love Got to Do with It?

The issue of the intersection between love and sex is also debated on gURL.com. Girls discussing this issue appear to be divided, as evidenced in the dialogue below:

Chelsea8194: *Um, sex isn't about love. i thought this was obvious, but i guess people don't know this. Sex is about reproduction. . . . i just don't think it's right to tell someone they have to wait until they're in love to start having sex. usually people have sex to feel good and because they like it.*
Justagirl: *sex is about love. It's how 2 people express their love to each other, it's not just something to do . . . if you have sex just to do it, it is wrong.*

Whether love is mentioned or not, for a few sexually active gURL.com members, sex can prove "satisfacting" for both parties:

OneMansKiss62: *my boyfriend and myself lost our virginities to each other. it was a day of pure passion and closeness. everything was satifacting and afterwards, i felt closer to him than ever*

Indeed, there can even be an element of *reciprocity* in sexual pleasure:

LuvinJustin: *the other night he was fingering me and he made me go crazy . . . i mean . . . it felt soo good. i was like shaking!! So, i wanna do suttin for him now that will make him go crazy like i did*

You Go, gURL

For some members of gURL.com, sexual pleasure can be something that does not require a partner at all. One of the topics of discussion on the "Shoutouts" board is masturbation. While much of the discourse around this topic concerns matters of technique, the two girls quoted below were exceptional in that they also addressed their own feelings of pleasure.

> dani8811: *i often hump a pillow because it feels really good . . . i'm too afraid to finger myself so i do that instead.*
> sickgirl1715: *if you sit on a pillow and think about something or someone and pretty much hump the pillow it works every time and you can do it a bunch of times and it feels good*

And when a gURL.com member is having trouble remembering that she, too, is entitled to her own sexual pleasure, there is often another gURL there to remind her:

> SnObOaRdBaBe311: *i have never given any guy head before . . . actually, i have never even made out or even kissed or ne thing (no, im not a goody good, i just haven't) . . . what am i supposed to do if he just whips it out . . . LOL . . . and expects me to suck it?*
> sarahmarie: *if he whips it out and expects you to suck it, just . . . say no. Just because he wants you to and is quite serious about it doesn't mean you are obligated to do it!*
> celloloony: *i think you should tell him to lick you, then when he says im done, say 'your not done' and make him do it some more then when he asks for head say you have to go*

In this dialogue, as in the three isolated comments below, we can see how the gURL.com community socially constructs norms for its members. While some gURLs abjure their sexuality, others are openly supportive of a girl's right to her own pleasure. Some girls are even willing to place female pleasure above that experienced by males:

> canadiangal103: *no matter what advice you get unless your really comfortable with doing it it won't be good. don't rush yourself. take your time and one night when your really into it things will happen and it will be the best you ever had.*
> cinna_bunz15: *if he cant deal with how u smell down there then he shouldn't be goin down on ne one—bottom line*
> Eyedream: *what i'm saying is you should do what feels right to you, not what feels right for him. THAT is girl power.*

What Are gURLs Talking about?

This question is more difficult to answer than it might appear. What struck us when we first began spending time on the pages of gURL.com—and what we continue to marvel at—is the diversity of the girls who have made this Web site a community. The girls of gURL.com vary in terms of age, attitude, worldview, and experience. We suspect that they represent a wide range of ethnic, socioeconomic, and geographic categories. We see clearly that they vary in their agency, self-awareness, and self-acceptance.

On the bulletin boards we visited, of course, what gURLs are talking about is sex. But not only sex. Yes, much of the discourse on the "Shoutouts" and "Dig or Dis" sex boards concerns sex acts. But more importantly, the girls who avail themselves of these boards are talking about, constructing, and performing *sexuality*. By "performing sexuality" we hew to the Butlerian (1990, 1993) conception of sexuality being performed—that is, enacted—in and through its verbal expression. The girls of gURL.com are doing more than merely describing their sexual experiences. In sharing their feelings and stories, they are bringing their sexual identities into being. And they are doing it in the context of a *community*—an all-female environment.

Yet we see a troubling paradox on gURL.com. The girls who enter this all-girl environment are limited in their ability to transcend the language and symbols of males. Into a community constructed and constituted by females, they bring androcentric conceptions of the distinct roles of men and women, attitudes and norms representative of a pre-Web world of male mastery and female submission. In gURL.com girls have a space to explore their budding sexuality—this is, to be sure, an innovation. But as is true of any community, there are limits. When girls stray too far, speak too explicitly, or express excessive agency, they are quickly reined in, reminded by other members of the community that certain larger societal expectations must still be met, even in this ostensibly progressive and "safe" space.

Of course, it is not gURL.com that is the originator of the proscriptions that the site's community enforces. The messages of patriarchy, androcentrism, and heteronormativity that girls receive—and send—in "real life" inevitably intrude into the constructed world of gURL.com. We assume that from their non-screen lives and their interactions with mass media, girls bring with them the sense of themselves as existing in relation to and even as subservient to males. Prior to, and concurrent with, their membership in the gURL community, they have learned that they are expected to subordinate not only their own pleasure but their very identity to males. Some gURLs even choose screen names that reify this relationality: How much agency can a girl claim when she identifies herself as "LuvinJustin," "jennylovesricky," or "adamschick4ever"? Giving oneself a "screen name" may be a late twentieth-century innovation, but women's practice of subsuming their own identities under those of men is timeless.

Another phenomenon we are troubled to observe in this otherwise new medium is the traditional valuation of sex. As their non-digital forebears have done for generations, the girls of gURL.com, too, recognize that sex is a marketable commodity. It is something they can offer that has exchange value, that can get them something: status, boyfriends, recognition as worldly. Girls' awareness that sex is a commodity overshadows, and may even make impossible, the view that sex can be something precious, sacred, or enjoyable. Even those girls who cling steadfastly to their virginity see it as something to be bartered in exchange for something else: security, reputation, love.

What is new, then? We suggest that contemporary "media scripts" (Christopher, 2001; Huesmann, 1986; Perse, 2001)—pornography, music videos, lyrics, and other mediated forms of communication—are important contributors to girls' conceptions of sexual behavior and identity. Today's scripts send mixed messages, but all encourage overt sexuality. Whether females are males' sexual objects (as in traditional movies and TV shows as well as contemporary music videos and video games) or even more agentive (as are the female characters in newer media texts such as *Buffy the Vampire Slayer* and *Sex and the City*), the predominant reading suggests soulless commodification. Even if, as Meyer and Shih (2003) assert, today's media portray women as procurers in the sexual marketplace, shopping for men as well as for designer shoes, they are still incomplete on their own, needing to fill a void with any branded acquisition.

Perhaps it is not surprising, then, that girls' own pleasure is addressed so rarely or so fleetingly. Attraction is acknowledged ("he is just so hot!"), but even then girls' own desire is sublimated to that of the boys in their lives. Sex for most gURLs is discussed as an obligation they must fulfill, either to their (male) partners or to their own sense of self-identity. Girls recognize that sex is expected of them; rarely do they allow themselves to believe or state that sex can be something *they* want for themselves.

Despite the many troubling, even depressing, revelations presented to us by gURL.com discourse, we do see glimmers of hope. That gURL.com is a *community* is heartening. Girls talk to each other and listen to each other; they share concerns, encourage each other, console each other, and let each other know that no one of them is going through her issues or feelings alone. Moreover, the community is constructed dialogically, as are girls' individual identities, in a forum that is generally safe.

The very *diversity* of the community is itself something positive. While gURL.com, like any community, has its boundaries, its members are not locked into one prescribed set of attitudes and behaviors about sex and sexuality. Beyond simply allowing girls to voice a wide range of opinions, feelings, and experiences, gURL.com is a place of (relative) tolerance: virginity, sexual activity, and expressions of heterosexual, homosexual, and bisexual attraction are all accommodated.

The discourses on gURL.com also fill an *information* gap. It is clear that many of the girls lack even the most basic knowledge about the mechanics of sex, reproduction and birth control processes, and the functions of their own bodies. Some of the older, more experienced members of the gURL.com community serve as educators, creating a forum where even the most naïve or ill-informed girls feel that they can ask questions and gain necessary knowledge safely, without risking censure or judgment for their innocence. The fact that girls feel that gURL.com is a place where they *can* talk about sexual issues is testament to the site's value.

Finally, while it is not true for the majority, some girls do honor their own capacity and desire for sexual pleasure. In freely and positively expressing their sexual feelings, they also serve as models for other gURLs, encouraging them to respect themselves and their own wishes, whether those wishes are to refrain from being sexual or to explore their sexuality or orientation. In sharing free, affirming self-expression, girls enact agency, demonstrating and performing their ability and their right to make their own choices; in other words, truly exerting girl/gURL power.

Notes

1. Our names appear in alphabetical order.
2. Data analyzed in the present paper were selected from postings appearing on gURL.com between October 15, 2003, and November 15, 2003. Having read all the postings in the "Shoutouts: Sex" and "Dig or Dis: Premarital Sex" sections during that period, we selected for analysis those which exemplified or illustrated any of a number of key themes, including sexual naïveté, promiscuity, homosexuality, fantasy, and sexual self-confidence or lack thereof.
3. Original spelling, capitalization, and syntax were retained in this and all other quotations hereafter. Ellipses in quotations indicate editorial elisions.
4. "gg me" is shorthand for "send me a gURL-gram." The Web site offers members the option of sending private instant messages, or gURL-grams. Some members choose to communicate one-on-one via gURL-grams rather than via the more public forum of the "Shoutouts" board.
5. It should be noted that sarahmarie, a regular contributor to discussions on the site, identified herself in an earlier post as a twenty-two-year-old peer counselor employed by Planned Parenthood.

References

Austin, J. L. (1962). *How to do things with words.* Cambridge, MA: Harvard University Press.

Bentley, M. K. (1999). The body of evidence: Dangerous intersections between development and culture in the lives of adolescent girls. In S. R. Mazzarella, & N. O. Pecora (Eds.), *Growing up girls: Popular culture and the construction of identity* (pp. 209–223). New York: Peter Lang.

Berger, P., & Luckmann, T. (1967). *The social construction of reality.* Garden City, NY: Doubleday.

Blumer, H. (1969). *Symbolic interactionism: Perspective and method.* Englewood Cliffs, NJ: Prentice-Hall

Butler, J. (1990). *Gender trouble: Feminism and the subversion of identity.* New York: Routledge.

Butler, J. (1993). *Bodies that matter: On the discursive limits of "sex."* New York: Routledge.

Christopher, F. S. (2001). *To dance the dance: A symbolic interactional exploration of premarital sexuality.* Mahwah, NJ: Lawrence Erlbaum.

Drill, E., McDonald, H., & Odes, R. (1999). *Deal with it!: A whole new approach to your body, brain and life as a gURL*. New York: Pocket Books.

Edut, O. (Ed.) (1998). *Adiós, Barbie: Young women write about body image and identity*. Seattle: Seal Press.

Gilligan, C. (1982). *In a different voice: Psychological theory and women's development*. Cambridge, MA: Harvard University Press.

gURL.com. (2003). What is gURL? Retrieved November 26, 2003, from www.gurl.com/more/about/whatisgurl.html

Heilman, E. E. (1998). The struggle for self. *Youth & Society, 30*, 182–208.

Holloway, D. L., & LeCompte, M. D. (2001). Becoming somebody: How arts programs support positive identity for middle-school girls. *Education & Urban Society, 33*, 388–408.

Huesmann, L. R. (1986). Psychological processes promoting the relation between exposure to media violence and aggressive behavior by the viewer. *Journal of Social Issues, 42*, 125–140.

Irvine, J. M. (Ed.) (1994). *Sexual cultures and the construction of adolescent identities*. Philadelphia: Temple University Press.

Ivinski, P. A. (2000, January/February). GURL power. *Print, 54*, 14.

Jain, A. (2003, August 6). iVillage makes first foray into teen space. *Crain's New York Business*. Retrieved November 27, 2003, from http://crainsny.com/news.cms?newsId=6196

Johnson, S. (1997). Theorizing language and masculinity: A feminist perspective. In S. Johnson, & U. H. Meinhof (Eds.), *Language and masculinity* (pp. 8–26). Oxford: Blackwell.

Joyce, E. (2001, May 29). dELia*s sells gURL.com to Primedia. *Internet News.com*. Retrieved November 28, 2003, from http://www.atnewyork.com/news/article.php/774441

Kroger, J. (1996). *Identity in adolescence: The balance between self and other*. London: Routledge.

Lakoff, R. (1975). *Language and woman's place*. New York: Harper & Row.

Lodge, S. (2000, January 31). Self-help for teens. *Publishers Weekly, 247*, 32–33.

Mantovani, G. (1996). *New communication environments: From everyday to virtual*. London: Taylor & Francis.

Mazzarella, S. R., & Pecora, N. O. (Eds.) (1999). *Growing up girls: Popular culture and the construction of identity*. New York: Peter Lang.

McRobbie, A. (1991). *Feminism and youth culture: From* Jackie *to* Just Seventeen. Boston: Unwin Hyman.

Mead, G. (1967). *Mind, self, and society: From the standpoint of a social behaviorist*. Chicago: University of Chicago Press.

Meyer, M. D. E., & Shih, Y. S. (2003, November). *So many roads, so many detours: Reading "Sex and the City" from a third-wave postmodern feminist perspective*. Paper presented at the annual convention of the National Communication Association, Miami Beach, FL.

Morrissey, B. (2003, August 5). iVillage acquires gURL.com. *Internet Advertising Report*. Retrieved November 27, 2003, from www.internetnews.com/IAR/article.php /2244631

Pecora, N. O., & Mazzarella, S. R. (1999). Introduction. In S. R. Mazzarella, & N. O. Pecora (Eds.), *Growing up girls: Popular culture and the construction of identity* (pp. 1–10). New York: Peter Lang.

Perse, E. M. (2001). *Media effects and society*. Mahwah, NJ: Lawrence Erlbaum.

Shandler, S. (1999). *Ophelia speaks: Adolescent girls write about their search for self*. New York: HarperPerennial.

Silverman-Watkins, L. T. (1983). Sex in the contemporary media. In J. Q. Maddock, G. Neubeck, & M. B. Sussman (Eds.), *Human sexuality and the family* (pp. 125–140). New York: Haworth.

Stern, S. R. (1999). Adolescent girls' expression on web home pages: Spirited, somber, and self-conscious sites. *Convergence, 5*, 22–41.

Stern, S. R. (2002). Virtually speaking: Girls' self-disclosure on the World Wide Web. *Women's Studies in Communication*, *25*, 223–253.

Sutton, M. J., Brown, J. D., Wilson, K. M., & Klein, J. D. (2002). Shaking the tree of knowledge for forbidden fruit: Adolescents learn about sexuality and contraception. In J. D. Brown, J. R. Steele, & K. Walsh-Childers (Eds.), *Sexual teens, sexual media: Investigating media's influence on adolescent sexuality* (pp. 25–55). Mahwah, NJ: Lawrence Erlbaum.

Turkle, S. (1995). *Life on the screen: Identity in the age of the Internet*. New York: Simon & Schuster.

Weiss, D., & Grisso, A. D. (2003, November). *What are girls talking about?: A content analysis of the adolescent girls' Web site gURL.com*. Paper presented at the annual convention of the National Communication Association, Miami Beach, FL.

Wyatt, C. A. (2000). Growing up girls: Book review. *Women's Studies*, *29*, 709.

Zaleski, J., & Gediman, P. (1999, July 26). Forecasts: Nonfiction. *Publishers Weekly*, *246*, 72.

easy task ... For many of us it seems that to be a feminist in the way we have seen or understood feminism is to conform to an identity and way of living that doesn't allow for individuality, complexity, or less than perfect personal histories.

Second, the ascension of Jammer Girls coincides with girls' access to the Internet. Unlike teen girl magazines that continue to emphasize beauty, consumerism, and passive interaction with the medium, the Internet allows girls to interact, interpret, and negotiate their world. For example, whereas magazines such as *Seventeen* emphasize music, "girls are provided with no information about how to set up a band, nor are they encouraged to learn to play an instrument" (Kearney, 1998, p. 291). They are hailed as consumers of music, not as makers of music. The World Wide Web, on the other hand, provides an avenue for girls to speak out about their refusal to accept "the established story of a woman's life" (Brown, 1991, p. 72). There are, for example, spaces for girls who are musicians to tell their stories to other girls who are eager to learn.

Web sites have a narrower focus compared with the broader view of other media. The Internet facilitates an intimate, transactional, informational relationship for young women, particularly in their search for knowledge about personal topics (Robbins, 2000). Girls' use of the Internet affords ways of negotiating social relationships and supported engagement in "personalized, self-directed and self-initiated learning" (Robbins, 2000, p. i). The accessibility and privacy of Web-based information offer girls personal and private space in which to explore questions about their bodies and their minds. Stern's (2000, p. 6) study of teen girl-created Web sites as public/private places reveals girls use them for "self-disclosure, especially self-clarification and self-expression." Dede (1996) explores girls' social life in the virtual world and describes how the Internet facilitates a social life of shared common joys as well as trials and tribulations. She notes the Internet provides a mechanism to maintain friendships while avoiding face-to-face contact. The Net is also a place where girls can enjoy a sense of freedom and a sense of control. Gruber (2003, p. 160) points out "Cyberspace is a place that allows for the 'complex and shifting play of body, self, and community.'"

Riot Grrls, paper zines, and bands such as Bikini Kill and Bratmobile preceded the Internet, but the compulsion to create them was, and remains, the same. Just as Riot Grrls did not "shy away from difficult issues and often addressed painful topics such as a rape and abuse" (Rosenberg & Garofalo, 1998, p. 810), Jammer Girls answer a similar call to action. Riot and other girl/gurl/grrrls are examples of "the independent, assertive, and empowering attitude of many young women who are not only entering previously male-dominated fields ... but are completely convinced they have every right to do so" (Denfeld, 1995, p. 135). Collectively, these pre-Web activities resulted in an "accumulation of practices" (Hawisher & Sullivan, 2003, p. 220) as second-wave feminists forged the way for other active feminist groups to extend their reach and claim virtual space. Just as girl zines were "often created to either resist or oppose representations of gender, sexuality, class, race, and age found in mainstream culture," today, several Internet sites "openly ridicule the dominant ideologies of female adolescence reproduced in mainstream girl magazines" (Kearney, 1998, p. 300) and, most particularly, the advertising.

While most Web sites provide net-savvy teens a heavy dose of popular culture (fan sites, fashion, and advice), others provide more pro-social content. The Web makes available space for activism, "for forging new social arrangements by creating a visual discourse that startles and disturbs" (Hawisher & Sullivan, 2003, p. 220), and it provides a place for women and girls to "act potently" (Haraway, 1991, p. 181). It is, however, up to girls to claim that space as their own.

Simultaneously, there has been "an escalation in the promotion of 'culture jamming' as a viable form of populist, anti-commercial critique" (Soar, 2002, p. 572). This activist strategy draws upon the ideas and ideals of the Situationists, a group who "first applied" the "spirit of anarchy to modern media culture" (Lasn, 1999, p. 100). Lampooning, critiquing, and parodying mainstream media representations are forms of such culture jamming. For example, *Adbusters*, the magazine and the Web site, is best known for spoof ads that aim "to subvert corporate brainwashing through satire and 'social marketing campaigns' such as 'Buy Nothing Day' and 'TV Turnoff Week'" (Sullum, 2000, ¶3).

"Subvertising" focuses on commercial messages and "uses the power of brand recognition and brand hegemony either against itself or to promote an unrelated value or idea" (Cortese, 1999, pp. 49–50). The intent is to turn the way we look at media "on its head" (Cortese 1999, p. 50) by actively appropriating "the ad's images . . . in critique and subversion" (Kearney, 1998, p. 300). The creation of culture jamming-focused Web sites, increased numbers of users, and gender-based user equity, have combined to extend the hands-on aspects of girl culture activism. The sister concepts of subvertising and culture jamming are strategies embraced by Jammer Girl activism and exemplified by the Web site About-Face.org.

Making an About-Face

The Web site About-Face.org, launched in 1997, encourages girls to voice their opinions, use the site as a "launching pad" for their own rebellion, and do so "with guts and humor" (Brunkala, 1998, ¶2). In 1997 and 1998, the site received an average of 3,580 page views per day, but by 2002, the hits were up to more than 11,000 (A-F.org). A-F.org uses a feminist pedagogical approach to guide young women through the process of "making an about-face, a 'reversal of standpoint,' when they look at advertising and media messages." The goal, according to its founders, is to get girls "to stop, turn around, and think about what they are seeing. Conversely, we agitate for an about-face in the way advertisers portray women" (A-F.org). The founding and ongoing charge of A-F.org is to critically address the hundreds of glossy advertisements contained in dozens of glossy magazines—ads that emphasize what girls and women should look like instead of what they are thinking and doing.

In the summer of 1995, Kathy Bruin saw, on the side of a bus, a Calvin Klein Obsession fragrance ad featuring a naked, reclining Kate Moss. Bruin and a friend found a print version of the advertisement, scanned it, *subvertised* (manipulated) it and, with help from friends and family, plastered posters onto temporary structures around San Francisco. In true Culture Jammer fashion, Bruin played on Moss' already thin body by exaggerating her skeletal features through enhancing shading and elongating the image in order to emphasize the model's already apparent vulnerability. She states that Kate Moss

> looked so young and gaunt, so frightened and vulnerable. This ad was the last straw for me. I wanted to make a statement that would be louder than just writing Calvin Klein a letter. I envisioned myself scrambling up scaffolding to deface billboards, or waiting at bus stops to attach an ad on the side of a bus! (A-F.org)

Bruin did not deface the ad, rather she satirized it by changing the text to "Emaciation Stinks!" and "Stop Starvation Imagery!" This led to the creation of the culture-jamming organization About-Face. In 1996, in honor of National Eating Disorders Week, Bruin and company created another poster "Bodies Are not Fashion Accessories: Question the Motives of the Diet Industry," and again displayed it around San Francisco, catching the attention of critics and comrades. A critique of the messages of fashion magazines' seasonal dictates of what is *in* and what is *out*, the poster satirized the objectification of women's bodies and presented the parts as baubles and beads. Bruin describes what motivated her to become an activist and why sexist advertising images are harmful to the psyches of both girls and boys:

> It does all of us a disservice to put such importance on the way women look rather than who we are. Encouraging women—and then expecting us—to spend a lot of time and money on beauty/fashion/diets guarantees that we may never measure up. Images of women in popular culture, led by the entertainment and "beauty" industries, affect women's perception of themselves and contribute to unhealthy relationships with food. This campaign encourages people to voice their own disenchantment and to make changes in their own lives. . . . If nothing else, we as individuals have the power to make an important impression in a young woman's life or to give a great big company a piece of our minds. (A-F.org)

Media Literacy

Based within the rubric of media literacy at the grassroots and online levels, About-Face.org believes that "when girls and women understand what media messages are saying and the impact of such messages, they will act to change them" (A-F.org). Indeed, this analysis of the strategies of A-F.org reveals the site practices the media literacy principle of "informed inquiry" (Center for Media Literacy, 2004), which involves a four-step process of awareness, reflection, analysis, and action through which young people acquire the skills to navigate the sea of media messages in order to:

- ¤ Access information from a variety of sources.
- ¤ Analyze and explore how messages are "constructed," whether print, verbal, visual or multi-media.
- ¤ Evaluate media's explicit and implicit messages against one's own ethical, moral and/or democratic principles.
- ¤ Express or create their own messages using a variety of media tools. (Center for Media Literacy, 2004)

In a nutshell, media literacy is "the ability to critically consume and create media" (New Mexico Media Literacy Project, 2004); the goal of which is to teach individuals how to decipher deeper meanings from the media they consume. A-F.org practices media literacy by applying the concepts to advertising as well as instructing girls how to use and transform ideas into their own words. It also serves as an exchange, a safe space to document action, post alternative responses, and create community. Specifically, a favorite trope that fits A-F.org well is the concept of *détournement*, defined as "an image, message or artifact lifted out of its context to create a new meaning" (Klein, 1999, p. 282), and which, drawing upon Debord (1983), is "a way for people to take back the spectacle that had kidnapped their lives" (Lasn, 1999, p. 103). This type of rebellion combines the typical teen emotions of angst and anxiety with the urge to rebel.

In 1996, the A-F.org staff began a media literacy program by giving speeches, talks, and workshops at colleges, high schools, and to women's organizations. A-F.org applies the same principles of awareness and critique to online examples, thereby reaching a wider geographical audience. The strategies and tactics give girls and women the tools needed to think critically about the advertising they encounter and provide an outlet for them to become active, educated, and visible consumers. Girls are encouraged to "use this Web site as a launching pad for [their] own rebellion" (A-F.org). The goal is to "disarm irresponsible advertisers" by subverting the messages that undermine girls' and women's strengths (A-F.org).

When interrogating A-F.org, it is important to ask whether girls actually use it to become more active and how they are actually "doing" activism there or elsewhere. In other words, does A-F.org foster and nurture Jammer Girls? In most cases it is difficult, if not impossible, to determine whether or not a particular "jam" was ultimately effective in changing corporate behavior or in motivating girls to do something because of visiting the Web site. However, A-F.org provides a space for Jammer Girls and school groups to post their information, art, and essays. Three links in particular demonstrate online Jammer Girl activism—"Your Letters," "Visitor Feedback," and "Your Forum" ("Visitor Picks," a fourth link, was not operational but is a site for submission of images).

Your letters

At the time of my research, twenty-four letters, and in some cases company responses to them, were posted on the "Your Letters to Companies" page. For example, there was an exchange between individuals and People for the Ethical Treatment of Animals (PETA) about the latter's sexist portrayals of women as a way of creating awareness for the plight of animals. Similarly, there were several letters about Skyy Vodka's sexist representations of women in their advertising. The first letter to Skyy, written in 2000, was posted, followed by a then-active link to a protester's Web page. Another letter, written in 2002, about a different Skyy ad, was also included, as was another link to a protester's page, along with information about how to contact the company and how to protest.

One example of change catalyzed by a letter written to a company concerned the "Fetish" perfume ads produced by Dana Perfume. The posted letter to Dana Perfume was signed by a group of concerned individuals, including a high school girl, an undergraduate psychology student, a psychology professor, a nurse, a business professor and "father of two daughters," and the associate director of a teen theater project. The objectionable advertisement, printed as an insert in the Portland *Oregonian,* shows a young woman wearing an orange bikini top, heavy pink eye makeup above and below her lashes, with a vial of Fetish perfume hanging between her breasts. The copy across her chest reads: "Fetish #16: Apply generously to your neck so he can smell the scent as you shake your head 'no.'" The group objected to what it felt was the ad's perpetuation of "the dangerous myth that girls and women mean 'yes' when they say 'no,'" and the resultant connotation that is it okay to ignore a girl's voice. Further, the ad was seen as reinforcing the stereotype that, by wearing certain kinds of clothing or makeup, a girl gives permission to be sexually violated. The text of the letter also mentions the rail-like thinness of the girl, how the red-rimmed eyes speak to the "heroin chic" look, and her overall battered look. The group asked the company to discontinue the ad, and said that until that time they would boycott Fetish products.

The response written by the general manager of Dana Perfumes referred to a telephone conversation she had with the psychology professor who signed the initial letter. The general manager wrote that the company stands behind the ads and does not believe them to be in bad taste or harmful to girls. In the end, however, she felt the group's letter demonstrated how the ads "do not live up to the standards that we attempt to maintain." Consequently, the ad campaign was discontinued.

Visitor feedback

Much like the guest books of many personal Web sites, A-F.org includes short quotes from visitors to the site, most of which exemplify visitors' excitement about and response to the site's content. At the time of my research there were nineteen posts. These posts provide evidence that A-F.org has been successful in contributing to girls' media literacy when it comes to images in advertising and, in some cases, actually spurred them on to activism. The following four posts exemplify the types of reactions included here:

- ¤ A.H. (Female, Age 18): *i loved it, it has to be the best thing i've foudn on the net in years!*
- ¤ J (Female, Age 18): *I absolutely love it. Never in my life have I felt as though someone really understood what I was going through as a size 12 in a highschool world of size 3s. As soon as I entered your site I felt immense relief . . . like I didn't have to do it alone anymore . . . like there was a whole world full of people out there who understood me. I can't find the words to express the immense gratitude I feel towards you. You have made me feel like I have worth even though I am 20 pounds over my "acceptable" weight based on height. I will be a full supporter of you forever, you've given me my life back, it's the least I can do in return.*
- ¤ V. (Female, Age 14): *I found you site very informing, and it was very nice to know that I am not the only one to think that some of the pictures in magazines are ridiculous, and out of control.*
- ¤ A. (Female, Age 14): *I loved this web site so much that I started writing letters and getting on the phone with companies. I even started my own webpage.*

In these cases, it is clear that A-F.org has been successful in creating a safe space and laying the foundations of a community for girls such as J, who does not feel "alone anymore," or V, who has discovered she is "not the only one" to think there are problems with these ads. Similarly, it evidences the site's role in empowering girls to actively resist and protest such messages; for example, A started writing letters, calling companies and "even started [her] own webpage."

Your Forum

In keeping with the site's overall role in providing a place in which girls can express themselves, the "Your Forum" link offers girls the opportunity to post original essays, art projects, student projects, reviews, and columns. Among the seventeen essays posted at the time of this research, thirteen focused on eating disorders and body image. Sixteen-year-old Katie Rainbow, for example, a high school student from Philadelphia, wrote: "I enjoy writing, reading anything I can get my hands on, and upsetting the natural order of the universe as often as possible." Katie uses chocolate as a metaphor in a brief essay in which she explores peer and parental pressure to be thin and popular, her disgust at having her cake and throwing it up too, increased awareness of what a life of denial means and, ultimately, her rebellion against a thin-obsessed culture. Similar to the poetry on girl-authored homepages, by using the third person in her essay, Katie is able to "address serious and intimate matters" yet remain one step removed from confessional writing (Stern, 2002, p. 277):

> She sees all the chocolate cake she wouldn't eat, and all the chocolate cake she threw up. And later, the chocolate cake she couldn't taste anymore. And the chocolate cake she couldn't swallow. And the chocolate cake she couldn't keep down. And now it's the chocolate cake that has been replaced with an IV tube in a horrible hospital where nurses yell at her when she doesn't eat her pudding. And they force that pound it took her so much work to loose back under her skin.

Another example of an art project is sixteen-year-old Erin's post of a collage of magazine images of unrealistically thin women in advertisements, fashion spreads, and art. The words "Shed the weight of the world" are placed on both sides, drawing the viewer's eyes back and forth across the piece. Similarly, Sarah submitted an image of a blond woman in a neon-pink string bikini with her wrists wrapped in rope, dangling like a puppet from a stick. Between the figure's arms are the words "The diet industry takes in over $40 billion each year and is still growing." To the right, she has typed "Don't participate in your exploitation."

Five student projects are linked to the "Forum" page including "Beauty is in the eye of the media," "Fashion: The cycle of shame," and the essay "Girl Power or Girl Downer?" written by fifteen-year-old Cate for her American Literature class. In this piece, Cate discusses Barbie, the Spice Girls, and cartoons as not only perpetuating the thin-is-in stereotype but also appropriating the concept of Girl Power by mainstream culture: "Having women in high positions. What is lie [sic] the media is feeding us? Can you say Girl Power?"

Conclusion

The purpose of this chapter was to explore the underpinnings of the false dichotomy of good girls and bad girls in American society and to posit the need for and appearance of a third teen girl identity. I argue that, because of the development of

Internet sites such as A-F.org, Jammer Girls have visibly emerged from the cloudy cultural waters of female adolescence. These sites have facilitated the concurrent "coming out" of nonconformist girls whose blossoming rose from soil made fertile by sister movements such as third-wave feminism. Other Web sites such as AdiosBarbie.com and Loveyourbody.org also strive to empower girls with information to work toward changing the flow of media images in order to help them love the bodies they are in and to value and voice their ideas. What A-F.org offers that is different from the others is a research-grounded media literacy framework within which images are analyzed, discussed, and culture jammed.

A shortcoming of this chapter and other feminist analyses of girl culture is that they primarily address the experiences of white, heterosexual, middle-class girls. There are vast and important opportunities within this research area to explore, for example, identity issues within other groups in American society, particularly among African American girls and Latinas. Social class issues also limit this kind of analysis. Not all girls have access to computers, and not all girls receive education in how to use them.

Has A-F.org made an impact on girls? Based on the anecdotal comments posted by girls visiting the site, the content of their essays, and the messages of their "jams," the answer is yes.

Negotiating the complexity of modern life as a girl is a full-time job. It has never been simple and it has not gotten any easier, as the playing field continues to be littered with complexities of celebrity culture, consumerism, and media ad-monishments such as makeup, clothing, and related accoutrements of female beauty. Peer pressure is as intense as ever, and the consequences of adolescent experimentation more serious than ever. If we as adults, parents, and educators can encourage girls, listen to their ideas, validate their feelings, the road can be less treacherous to traverse. Internet sites such as About-Face.org, media literacy education in schools and on the Internet, girls' participation in sports, and the burgeoning critical mass of girls who are, to paraphrase the film *Network*, "mad as hell and not taking it anymore," all offer hope that girls of the future will be in a better position to choose their identities rather than finding them imposed. In conclusion, A-F.org founder Kathy Bruin offers this positive message:

> We must all choose between battles: One battle is against the cultural ideal, and the other is against ourselves. Must we always define ourselves by what popular culture dictates? Develop your own style. Have fun—wear lipstick. Or don't. You're the boss of you. By speaking out and accepting yourself (dimples and all), you help break the barriers. (A-F.org)

Note

1. The expression "culture jamming" was first used by the jammer band Negativland (http://www.negativland.com/) and was popularized by Dery (1993), who defines it as "media hacking, information warfare, terror-art, and guerrilla semiotics all in one" (http://www.levity.com/markdery/culturjam.html). The expression has since become associated with Kalle Lasn and Canadian activist group Adbusters.

References

About-Face.org. Retrieved December 02, 2003, from http://www.about-face.org

Avery, R. K. (1979). Adolescents' use of the media. *American Behavioral Scientist, 23*, 53–70.

Boyle, G. (2004). Girlfighting. Retrieved February 5, 2004, from http://www.colby.edu/colby.mag/issues/win04/media

Brown, L. M. (1991). Telling a girl's life: Self-authorization as a form of resistance. In C. Gilligan, A. G. Rogers & D. L. Tolman (Eds.), *Women, girls, and psychotherapy: Reframing resistance* (pp. 71–86). New York: Harrington Park.

Brown, L. M. (2003). *Girlfighting: Betrayal and rejection among girls*. New York: New York University Press.

Brumberg, J. J. (2000, November 24). When girls talk: What it reveals about them and us. *Chronicle of Higher Education, 47*. Retrieved November 11, 2003, from http://chronicle.com/cgi2-bin/texis/chronicle/search

Brunkala, K. (1998, November 10). Fighting bad images. *Cleveland Plain Dealer*. Retrieved November 15, 2002, from Lexis-Nexis.

Bucholtz, M. (1999). Bad examples: Transgression and progress in language and gender studies. In M. Bucholtz, A. C. Liang & L. A. Sutton (Eds.), *Reinventing identities: The gendered self in discourse* (pp. 3–24). New York: Oxford University Press.

Caron, A. (2000). Parenting adolescents. Retrieved February 8, 2004, from http://www.anncaron.com/column.cfm

Center for Media Literacy. Retrieved February 15, 2004, from http://www.cml.org

Coleman, J. S. (1961). *The adolescent society*. Glencoe, IL: The Free Press.

Cortese, A. (1999). *Provocateur: Images of women and minorities in advertising*. New York: Rowman & Littlefield Publishers.

Cusimano, D. C., & Thompson, J. K. (1997). Body image and body shape ideals in magazines: Exposure, awareness and internalization. *Sex Roles, 37*, 701–721.

Debord, G. (1983). *Society of the spectacle*. Detroit, MI: Black & Red.

Dede, C. (1996). Emerging technologies and distributed learning. *American Journal of Distance Education, 10*, 4–36.

Denfeld, R. (1995). *The new Victorians: A young woman's challenge to the old feminist order*. St. Leonards, Australia: Allen and Unwin.

Dery, M. (1993). Culture jamming: Hacking, slashing and sniping in the empire of signs. Retrieved January 15, 2004, from http://www.levity.com/markdery/culturjam.html

Driscoll, C. (2002). *Girls: Feminine adolescence in popular culture and cultural theory*. New York: Columbia University Press.

Erikson, E. (1950). *Childhood & society*. New York: W.W. Norton.

Field, A. E., Cheung, L., Wolf, A. M., Herzog, D. B., Gortmaker, S. L., & Colditz, G. A. (1999). Exposure to mass media and weight concerns among girls. *Journal of Pediatrics, 103*, E6.

Gilligan, C. (1982). *In a different voice: Psychological theory and women's development.* Cambridge, MA: Harvard University Press.

Gruber, S. (2003). Challenges to cyberfeminism: Voices, contradictions, and identity constructions. In L. Gray-Rosendale, & G. Harootunian (Eds.), *Fractured Feminisms: Rhetoric, context, and contestation* (pp. 159–176). Albany, NY: State University of New York Press.

Hall, S. (1980). Encoding/decoding. In S. Hall, S. Baron, M. Denning, D. Hobson, A. Lowe, & P. Willis (Eds.), *Culture, media, language* (pp. 128–138). London: Hutchinson.

Haraway, D. J. (1991). *Simians, cyborgs, and women: The reinvention of nature.* New York: Routledge.

Hargreaves, D. A., & Tiggemann, M. (2003). Female "thin ideal" media images and boys' attitudes toward girls. *Sex Roles, 49*, 539–544.

Harrison, K. (2000). The body electric: Thin-ideal media and eating disorders in adolescents. *Journal of Communication, 50*, 119–143.

Harrison, K., & Cantor, J. (1997). The relationship between media consumption and eating disorders. *Journal of Communication, 47*, 40–67.

Hawisher, G. E. & Sullivan, P. (2003). Feminist cyborgs live on the World Wide Web: International and not so international contexts. In M. E. Hocks & M. R. Kendrick (Eds.), *Eloquent images: Word and image in the age of new media* (pp. 220–235). Cambridge, MA: MIT Press.

Hofschire, L. J., & Greenberg, B. S. (2002). Media's impact on adolescents' body dissatisfaction. In J. D. Brown, J. R. Steele & K. Walsh-Childers (Eds.), *Sexual teens, sexual media: Investigating media's influences on adolescent sexuality* (pp. 125–149). Mahwah, NJ: Lawrence Erlbaum.

Jennings-Walstedt, J., Geis, F., & Brown, V. (1980). Influence of television commercials on women's self-confidence and independent judgment. *Journal of Personality and Social Psychology, 38*, 203–210.

Kearney, M. C. (1998). Producing girls: Rethinking the study of female youth culture. In S. A. Inness (Ed.), *Delinquents & debutantes: Twentieth century American girls' cultures* (pp. 285–310). New York: New York University Press.

Kilbourne, J. (1999). *Can't buy my love: How advertising changes the way we think and feel.* New York: Simon & Schuster.

Klein, N. (1999). *No logo: Taking aim at the brand bullies.* New York: Picador.

Lashbrook, J. T. (2000). Fitting in: The emotional dimension of adolescent peer pressure. *Adolescence, 35*, 747–757.

Lasn, K. (1999). *Culture jam: The uncooling of America.* New York: William Morrow.

Lee, M. (2003). *Fashion victim: Our love-hate relationship with dressing, shopping, and the cost of style.* New York: Broadway Books.

Levine, M. P., Smolak, L., & Hayden, H. (1994). The relation of socio-cultural factors to eating attitudes and behaviors among middle school girls. *Journal of Early Adolescence, 14*, 471–490.

Lowe, M. (2003, June). Colliding feminisms: Britney Spears, "tweens," and the politics of reception. *Popular Music and Society, 25*, 124–140.

Lueker-Harrington, D. (2001, July 10). Marketers. *USA Today*, p. 13A.

Marcotte, D., Fortin, L. Potvin, P., & Papillon, M. (2002). Gender differences in depressive symptoms during adolescence: Role of gender-typed characteristics, self-esteem, body image, stressful life events, and pubertal status. *Journal of Emotional and Behavioral Disorders, 10*. Retrieved December 05, 2003, from Ebsco Host.

McCabe, M. P. (2001). The structure of the perceived sociocultural influences on body image and body change questionnaire. *International Journal of Behavioral Medicine, 8*. Retrieved September 16, 2003, from http://epnet.com

McRobbie, A. (2000). *Feminism and youth culture* (2nd ed.). New York: Routledge.

New Mexico Media Literacy Project. Retrieved February 28, 2004, from http://www.nmmlp.org/medialiteracy.htm

Offer, D., Schonert-Reichl, K. A., & Boxer, A. M. (1996). Normal adolescent development: Empirical research findings. In M. Lewis (Ed.), *Child and adolescent psychiatry: A comprehensive textbook* (pp. 280–290). Baltimore: Williams & Wilkins.

Quart, A. (2003). *Branded: The buying and selling of teenagers*. Cambridge, MA: Perseus.

Robbins, J. I. (2000). *Making Connections: Adolescent girls' use of the Internet*. Unpublished doctoral dissertation, Virginia Tech.

Rosenberg, J., & Garofalo, G. (1998). Riot grrrl: Revolutions from within. *Signs: Journal of women in culture and society, 23*(3), 809–841.

Shields, V. R. (2002). *Measuring up: How advertising affects self-image*. Philadelphia: University of Pennsylvania Press.

Soar, M. (2002). The first things first manifesto and the politics of culture jamming: Towards a cultural economy of graphic design and advertising. *Cultural Studies, 16*(4), 570–592.

Stern, S. (2000). Adolescent girls' home pages as sites for sexual self-statement. *SIECUS Report, 28*, 6–15.

Stern, S. R. (2002). Sexual selves on the World Wide Web: Adolescent girls' home pages as sites for sexual self-expression. In J. D. Brown, J. R. Steele, & K. Walsh-Childers (Eds.), *Sexual teens, sexual media: Investigating media's influence on adolescent sexuality* (pp. 265–285). Mahwah, NJ: Lawrence Erlbaum.

Sullum, J. (2000, May). Rebel without applause. *Reason*. Retrieved December 29, 2003, from www.findarticles.com

Sutton, L. A. (1999). All media are created equal: Do-it-yourself identity in alternative publishing. In M. Bucholtz, A.C. Liang, & L.A. Sutton (Eds.), *Reinventing identities: The gendered self in discourse* (pp. 163–180). New York: Oxford University Press.

Tan, A. (1977). TV beauty ads and role expectations of adolescent female viewers. *Journalism Quarterly, 56*, 283–288.

That's me! Retrieved February 12, 2004, from http://www.netw.com/~abbaida/musings/thatsme.htm

Wolf, N. (1991). *The beauty myth*. New York: William Morrow.

4

Gender, Power, and Social Interaction

How Blue Jean Online Constructs Adolescent Girlhood

Susan F. Walsh

Introduction

Millions of adolescent girls rely on popular teen magazines such as *Seventeen*, *YM*, and *Sassy* for information about fashion, beauty, dating, sex, and relationships (Duffy & Gotcher, 1996; Duke & Kreshel, 1998; Durham, 1998; Peirce, 1990). Findings from studies of advertising and editorial content reveal that these magazines often portray girls as the object of male desire, obsessed with appearance and romance (Carpenter, 1998; Duffy & Gotcher, 1996; Garner, Sterk, & Adams, 1998; Lamb, 1999; Markham, 1993; Merskin & Silva, 2000). Girls also rely heavily on articles featuring the opinions of boys to formulate their own feminine identities (Duke & Kreshel, 1998; Mazzarella, 1999).

Researchers have also examined the extent to which mainstream girl's magazines have appeared to support feminism, such as when a feminist subtext has been detected in content (Budgeon & Currie, 1995).[1] The results indicate that editors and advertisers consistently present girls with a narrow construct for femininity, while at the same time portraying them as independent and autonomous, thereby reinforcing both traditional and modern aspects of femininity. Advertising content in particular remains a prominent source for the co-option of feminist ideals. Although the context within advertising appears to have shifted in recent years to one that underscores a female's ability for personal (and sexual) freedom and self-definition, the construct of femininity continues to be framed in terms of fashion and beauty, thereby making feminism less about social and political ideals and more about one's individual lifestyle choices (Budgeon, 1994; Goldman, 1992).

Similar contradictions occur in teen magazines where readers are presented with messages that criticize gender inequality and, at the same time, are offered pleasurable or simple resolutions to the tensions it causes (Budgeon & Currie, 1995). *Seventeen*, for example, instructs readers to be assertive and independent and to avoid pressure from boys to engage in activities that make them feel uncomfortable—but only if they have not yet had sexual relations (Budgeon & Currie, 1995). Girls who *have* engaged in intercourse are advised to assume responsibility for monitoring and correcting any sexual problems that might occur in the relationship such as waning male sexual desire and impotency (Garner, Sterk & Adams, 1998).[2] Durham (1998) describes the way girls' sexuality is negatively constructed in *Seventeen* in terms of "discontinuities" that, on one hand, urge girls to be sexually desirable, and on the other, discourage them from engaging in sexual activity (p. 383). As Goldman (1992) has noted, this postfeminist view constructs the female body paradoxically as a site where antagonism exists between control over one's body and patriarchal assumptions about home, love, and sex. It also presents female sexuality as a commodity that is then re-presented to readers in the form of expressions of individual freedom and subjectivity.

These studies demonstrate how traditional girls' magazines have reinforced narrow and stereotypical meanings about girlhood as preferred readings in advertising and editorial content, often while simultaneously appropriating a feminist discourse. The commodification of femininity and female sexuality presents readers with a very specific construct linking power, control, and individualism to material consumption that gets naturalized and carried into girls' social interactions (Currie, 1999; Mazzarella, 1999). Though often characterized as "alternatives," these portrayals do not offer readers anything new or different in terms of stereotypes of women (Budgeon, 1994). Moreover, as Durham (1998) has argued, such constructions serve to "sustain and support the social power dynamics that keep girls sexually subordinated and contained" (p. 386).

Further support for the study of girl-zines can be found in the historical relationship between traditional girls' magazines and advertising. Print magazines rely heavily on advertising revenue, placing pressure on editors to avoid coverage of certain potentially controversial issues (Steinem, 1990). The reason so few alternative girls' magazines exist in print is most likely due to the inherent difficulty editors have in maintaining a balance between advertisers' interests and (potentially) provocative content. Because online magazines can be relatively inexpensive to produce, they are less reliant on advertising revenue (Chonin, 1999). The resulting lack of pressure to conform to the status quo might lead to an increase in feminist content aimed at adolescent and teen girls.

The research presented thus far indicates that girl culture remains a fruitful area for scholarship relating to the social construction of gender, with much of the attention (rightly) focused on girls as passive consumers and girls' magazines as sources of female oppression. Few studies, however, have considered girls as "active meaning makers" (Stern, 2002, p. 227). Feminist scholars have recently begun to recognize cultural alternatives to mainstream media, including the Internet, as opportunities for girls to gain access to active voice, suggesting that those who seek to understand the current state of girlhood should expand their investigations to include cultural texts produced *by* girls (Kearney, 1998).

Reading the Alternative

According to the Center for Media Education (CME) (2001), the majority of teens in the U.S. choose Internet technology over any other medium, including television, as their preferred source for information. No longer the purview of an elite group consisting mainly of academics, the Internet is a central part of American daily life at home, work, and school. The CME study also found that nearly three-quarters of teens between the ages of twelve and seventeen are visiting Web sites, the majority of which revolve around popular culture. These findings illuminate the need to gather information about teens and the World Wide Web—presently the domain of market researchers—in order to better understand the impact this technology may have on youth in general and girls in particular.

A wide variety of online magazines aimed at girls between the ages eleven and twenty currently exist on the World Wide Web, ranging from those that are corporately financed to a single person's homemade creation (Chonin, 1999; Rowe, 1997). Many girl-zines are gendered in the traditional manner outlined above (Center for Media Education, 2001). Several, however, do provide alternatives to the beauty-glamour construct by giving girls access to more substantive information about more diverse topics, including sports, travel, health, education, careers, community service, and feminism (Walsh, 2002). Given that this relatively new medium has thus far avoided serious scrutiny by feminist researchers, studies of the ways in which it contributes to the social construction of gender seem timely and worthwhile.

Beginning with the assumption that females are devalued, feminist scholars examine societal institutions, including the media, to determine how this ideological perspective gets represented at any given point in time (Jansen, 2002; Steeves, 1987). However, the liberal feminist assumption that women will be perceived as more empowered, or as powerful or capable as men if only they are employed with equal status in media organizations or portrayed more often in media content, dismisses the balance between the achievement of individuals' goals, desires, and interests and the intricate web of social contexts and relationships from which they originate. This study is, therefore, grounded in the belief that critical feminist theory can and should function as both a critique of contemporary society (Jaggar, 1988) as well as a starting point of inquiry for discovering the means for social and political change (Harding, 1991).

A second theoretical assumption shaping the study centers on the way in which subjectivity functions in cultural texts as a site for the construction of social meaning. As Currie (1999) has noted, subjectivity is achieved through expressions of power, agency, and social and cultural life that combine to create meaning. Girls who read traditional teen magazines do engage in a form of "meaning-making," but they do not create the conditions under which that process occurs, which keeps them positioned as objects, or "embodied female subjects" within those texts (p. 309). Because they are produced by, for, and about girls, Internet sites present a unique opportunity for the existence of various subjectivities and their associated expressions of power and agency.

Also relevant to this research is Gramsci's (1971) notion of hegemony—referring to the manner in which dominant ideologies become invisible when they are translated into commonsense understandings of the way societies work. Readers of traditional girls' magazines, for example, are exposed to an ideological construct that links femininity to attractiveness and consumerism. Although individuals who occupy the same ideological system will likely construct preferred or dominant meanings (Hall, 1980), hegemonic ideologies may also be opposed or resisted. Moreover, the encoding and decoding processes that occur in the production and reception of media texts are inextricably linked, making negotiations over meanings and values more likely to occur (Fiske, 1989; van Zoonen, 1994).

Studies of popular culture continue to be valuable tools for noting ideological changes and trends in society. Girl-zines, in particular, provide media researchers with an opportunity to determine the extent to which and how dominant ideologies regarding gender and power are challenged as a result of second- and third-wave feminism. To that end, I have chosen to focus on Blue Jean Online because it appears to have the potential to be dramatically different than traditional print magazines for teen girls.[3] Described by *USA Today* as "the thinking girl's magazine," it has received endorsements from a variety of sources including the Feminist Majority Foundation, the American Association of University Women, *Ms.*, and the Ms. Foundation for Women.[4] The all-female staff consists of the publisher, twenty editors, correspondents, illustrators, and writers between the ages of fourteen and twenty-four years.[5] They, along with the audience, create the content for the Web site.

This study consists of a qualitative interpretative analysis, or a "deep read" of the Web site over a twelve-month period.[6] During that time, I attempted to identify any overarching themes that appeared to challenge the dominant discourse relating to gender roles and set out to determine the extent to which and how feminist perspectives were being presented. As van Zoonen (1994) has noted, using both interpretative and qualitative approaches provides a broader framework for feminist researchers who set out to analyze the process of gender construction as a product of power relations and social interactions.

Challenging Norms: Gender Roles, Sexuality, and Political Activism

Linda Steiner (1995) has argued that critiques of so-called "women's media" must confront the dominant discourse of both womanhood and feminism. In a thoughtful essay, she proposes an admittedly "ideal" set of criteria for media *truly* wishing to "stand up" for women—and, I would add, girls—that includes giving females expression and control over their media, empowering their lives, celebrating their struggles and successes, respecting their intelligence, recognizing their diversity, and avoiding practices that harm their interests. On this last point, Steiner challenges the media to establish policies and practices that prevent the objectification of women (and girls)—to serve them physically, emotionally, and intellectually—and to stop linking female equality and status to the purchase of products and services.

When looking at Blue Jean Online (BJO), Steiner's principles do not appear to be as idealistic as they once seemed. Indeed, unlike traditional teen magazines that objectify femininity, Blue Jean Online presents girls and young women as autonomous and subjective beings, a philosophy made explicit in the magazine's mission statement:

> Welcome to Blue Jean Online—the only Web site written and produced by young women from around the world. Over 1,000,000 visitors from 100 countries have visited us to see what young women from around the world are thinking, saying and doing everywhere. Blue Jean Online is a creative space for young women ages fourteen to twenty-two to submit their writings, art work, photography, crafts and other works for online publication to a world wide audience. Our young women editors and correspondents are the driving force of Blue Jean Online. This coalition of diverse young women is dedicated to publishing what young women are thinking, saying, and doing. We profile girls and women who are changing the world!

Support for Steiner's perspectives was present in numerous articles, including "Growing up Female: A World of Contradictions," in which a young woman questions patriarchal assumptions about body image, self-concept, and the construct of femininity.

A young woman growing up in America faces several complexities and contradictions. She clings to her childhood like a security blanket while trudging ahead into the uncertainties of adulthood. She bursts with energy, all the promises of the future gleaming in her eyes, but she is pulled back far too easily by her own fear. She's told by society that innocence is sexy, but that sexiness is dangerous; that brains are a necessity but brainpower is defeating. She stands in front of a mirror for hours trying to figure out who she really is while never realizing that she's looking in all the wrong places.

Similar admonishments of society's role in promoting the sex and beauty construct that objectifies females were found in "Can't Buy My Love: Jean Kilbourne Takes on the World of Advertising,"[7] "Girls Take on Hollywood by Declaring 'Turn Beauty Inside Out Day,'" and "Porn Star or Pop Star?" In the latter, Britney Spears and Christina Aguilera are criticized for "[getting] paid big bucks to shed their clothing and instigate sexual fantasies."

It shouldn't be acceptable—end of story. Yet while the music business needs an overhaul, I realize it won't happen overnight. What it needs to start that overhaul, though, is a strong reaction from people like you. Changing the channel is not going to be enough anymore—not while a ten-year-old girl in another state will still be watching. . . . Remind the young people in your life that pop music videos are not reality. . . . Inspire others to be participants when watching television, not passive spectators.

As was noted earlier, the commodification of female sexuality in popular culture must be considered in relation to capitalism and patriarchy in order to understand its function in the production of meaning. In a feminist critique of Jewel's music, a young woman expresses opposition to this hegemonic notion that necessarily keeps female sexuality, power, and success tied to materialism:

Yes, the same woman who defends her songs as touting anti-commercialism messages actually sells out . . . to a new razorblade. Jewel was defending her "Intuition" music video's sexualized images as parodies of commercialism, but meanwhile she was profiting from commercialism. The irony of this left me shaking my head in front of the television. My hopes for the forthcoming *0304* were disappearing as quickly as Jewel's devotion to her principles.

Further critiques of the body/body image construct appeared in book reviews such as this one written by a fourteen-year-old correspondent about a popular young actress with an eating disorder:

Room to Grow: An Appetite for Life by Tracey Gold is about her life and her battle with anorexia, but if you're looking for preachy, "look-at-poor-beautiful-me," tear glistening in eye reading, then you won't find it here. . . . Like most little girls, Gold had grown up wanting to please and (with a little help from movies and books) began to believe that if she was skinnier, life would be better. The book itself is not so much autobiographical as it is informative. Gold's life acts simply as a backdrop to the main topic, a disease that kills.

BJO meets another of Steiner's criteria, which is to respect women's diversity and intelligence by addressing provocative topics and complex issues. For example, stereotypes about masculinity and male homosexuality were explored in several articles, including "Redefining Manhood in the 21ˢᵗ Century," in which the author discusses survey results suggesting that men are beginning to explore their feelings without fear of losing their status as "real" men. "Men shouldn't have to risk their masculinity when a movie makes [them] cry. The media may continue to question that but in the end . . . more men will respond with a perspective that is, thankfully, quite different from traditional expectations." Stereotypical assumptions about homosexual men are similarly criticized in a review of the teen novel *Boy Meets Boy*, in which the author contrasts portrayals of gays in the book with those on popular televisions shows like *Will and Grace* and *Queer Eye for the Straight Guy*. Pointing to what she describes as a "unique irony," she writes:

> Today's media . . . accepts homosexuality but permits homosexuality only by perpetuating stereotypes. Would there really be interest in *Queer Eye for the Straight Guy* if it didn't exploit the stereotype of gay men dressing well? . . . A new book however, has set out to challenge these stereotypes and double standards—by not mentioning them.

It is important to note here that the mere presence of discourse relating to males on a feminist Web site is a far less significant factor than the actual meanings about masculinity that are being constructed in a girls' magazine. Whereas males are typically portrayed in traditional teen magazines as independent and active (and girls as weak, dependent, and passive), contributors to BJO present an alternative discourse portraying dominant stereotypes of the stoic heterosexual male or the inherently stylish gay man as limiting and restrictive. Moreover, the magazine serves as a site where oppositional readings about male and female subjectivities are constructed in ways that challenge a hegemonic process that keeps white heterosexual males largely in control of the popular culture industry.

Steiner's assertion that women and girls should control their own media was reflected in several feminist critiques of dominant masculinity, including an exchange among editors, correspondents and readers concerning whether or not the magazine should begin reviewing "boy bands." Although several girls did argue in favor of the move—mostly out of concern that the magazine might be perceived as anti-male—the majority of responses contained expressions of disapproval, primarily on the basis that doing so would violate BJO's mission. The following statements exemplify the extent to which girls felt betrayed by suggestions for changes that would in any way give males primary subjectivity or prioritize them within the magazine.

> I do NOT think that we should review boy bands. . . . BJO is a young women's space, and it's not exclusive or feminazi to close-minded [sic] to stick to our purpose.

> I enjoy coming to this website because I get to hear about other girls and what's going on with them on our side of the world. If I wanted boy band reviews, I'd go to *Teen People*. . . .

I'm a girl rocker who loves hearing about girl bands and the latest feminine issues.

Blue Jean Online meets another of Steiner's objectives for "real" girls' and women's media by shifting its focus away from self-improvement, while recognizing that many problems cannot be solved simply by changing one's attitude or behavior—essentially, that the oppression of females is the fault of society, not individuals, and they must act collectively on their own behalf to end it. In one article, for example, a young woman criticizes the inappropriate behavior of males at concerts, including band members who encourage such behavior.

> I've got a problem. Specifically, I've got a problem with drunk boys showing up at concerts, yelling about how they want to get on stage to kiss and grope my favorite rock stars, and elbowing me in the ribs. I'm tired of boys who throw beer at me when I ask them to knock it off. I'm sick of guys who apparently think girls go to see a rock show to get felt up, pushed around, and beat up. . . . I'm disappointed by bands who seem to encourage these guys . . . and I'm tired of complaining about it. [Women need to] show a little solidarity. If the girl next to you is getting pushed around, let her know you're sympathetic. Band together with the fans surrounding you to form a bully watch, and shove the jerks to the back of the crowd where they belong.

Sexual harassment and assault are addressed elsewhere on the Web site. An essay titled "Hooters Air" questions the restaurant chain's proposed foray into the airline industry, while another article draws attention to the importance of taking self-defense courses.

> Someone is being raped as you read this. According to the U.S. Department of Justice, a woman is sexually assaulted somewhere in America every two minutes. Hundreds of thousands of women are raped or assaulted every year. One in two rape victims is under the age of 18, and teenagers between the ages of 16 and 19 are three and a half times more likely to be victims of rape, attempted rape, or sexual assault. Most of us have heard these statistics, or ones like them. We live in a society that has a problem: many men think it's okay to be violent towards women. Our society also teaches us from a very young age that women are convenient targets for male aggression because they can't fight back. Well, that just isn't true. I am standing in a line of a small group of courageous women. We have all decided that we need to say "NO!" "No" to a society that tells us we cannot defend ourselves, and no to a society that teaches us to be polite, to smile needlessly, to risk the threat of violence in order to avoid "making a scene." We have decided that our safety is important.

Other articles address the oppression of girls and women globally, including "Tales of Veils," a pictorial essay on western stereotypes of Emarati women who cover themselves. The following letter is from a reader responding to the article, "A Girls Fight Against Dowry in India."

> I loved the article . . . I've read a book about that topic, called *Homeless Bird*, by Gloria Whelan, which is about a thirteen-year-old girl, Koly, who must learn to fend for herself after

her husband of only a few days dies, and she is left in India's "city of widows." . . . Since then, I've really been interested in India, and I'm glad that the dowry's now illegal.

Reflections of Steiner's call for the celebration of women's collective struggle and success occur at many levels within the magazine, from book reviews ("Where the Action Was: Women War Correspondents in World) to a (male) reader's response to an article about gender equality:

I can't agree more with [the editor's] closing comment, that the problem isn't men holding the door for women, but rather the one-sided nature of it all. Women should do the same for men when given the chance. This means it isn't so much traditional one-sided chivalry but rather something called "common courtesy."

A regular feature titled "Women We Love" profiles a wide range of unique and interesting females.[8] An important by-product of this particular feature is the dialogue it often generates, such as when readers, editors, and correspondents reacted angrily to letters from adults questioning the "appropriateness" of an article about Eve Ensler and *The Vagina Monologues* as seen in the following five examples:[9]

The truth has been universally acknowledged as inappropriate and much too strong for young girls to handle. Of course, not many things universally acknowledged are true. I find it ridiculous—and yes, I've chosen the perfect word to describe it—that anybody should find Blue Jean or the *Vagina Monologues* piece offensive. This stems from the same idea that "if I ignore it, it will surely go away because it can't happen to anyone I know." Blue Jean is about empowering girls and women. How do you empower people? Information, as we should all know is power.

. . . Teenage girls are sooner or later going to grow up into mature adults. I feel it is better for them to come in contact with the real world now rather than later, when it might be too late.

One of the immediate things I noticed about Blue Jean Online which eventually influenced me to pursue a position as a correspondent with it was its core mission—a place for young women to go and feel comfortable. It was not, and continues to be [sic] a place where women can express themselves without conforming to standards and not be harshly edited or restricted, though it is all done in a tasteful manner. . . . *The Vagina Monologues* tackle [sic] much more important issues than sexuality . . . [It] deals with worldwide problems from domestic violence to rape.

I think this is one of those situations where you have to say 'whatever.' If these folks are "offended" by [the] piece on *The Vagina Monologues*, I would question their level of understanding of it (if they had even ever seen or heard of it previously). The suggestion that [we] should involve parents and educators to steer BJO content is interesting. . . . I think we can all agree this is not the founding ideology that Blue Jean Online functions under. On the whole, this series of email communications between this parent and the principal strikes me as entirely uninformed, knee-jerk conservative, and ignorant of the importance of Eve Ensler's progressive and creative work.

These emails reinforce societal fears of educating girls about their blossoming sexuality. Ignorant parents and educators fear that the mere mention of words like vagina, pregnancy, or rape will make way for girls' rampant promiscuity. By refusing to teach girls about their bodies, all society does is produce girls who are unprepared to make strong decisions.

Taken as a whole, these five examples suggest that Blue Jean Online might contribute significantly to girls' socialization by providing them with constructions of alternative meanings about femininity, as well as the rhetorical spaces necessary for constructing their own feminist identities.

Symbols of resistance to institutionalized patriarchy and oppression were present in several BJO articles about environmental, political, and social activism, reflecting Steiner's contention that "real" women's magazines take risks by tackling weighty issues—even at the risk of upsetting readers. In "Hamburger Politics," for example, a correspondent draws attention to the ways in which cattle ranching practices in Central America have devastated rainforests and the role that U.S. consumers and industries have played in promoting the destruction:

> Large areas of rainforest are being destroyed simply so that some of us can eat this beef. . . . The rainforests are vital to the global environment. Three-quarters of the world's species make their homes there, and these forests produce much of the world's oxygen supply and provide water for over one billion individuals. If such cattle ranching practices are allowed to continue, all of this may vanish.

An article titled "Chicken Farm Rescue Mission," offers a similarly graphic account of the cruel and inhumane conditions that exist within the commercial egg industry. The author concludes with an emotional appeal to readers that they become more aware of the need for animal rights activism:

> You may never have the opportunity to rescue birds from a factory farm. However, you can help stop the exploitation of chickens by eliminating, or at least limiting eggs (and other animal products) from your diet. You can also help by learning more about the animal exploitation in factory farms, and sharing what you learn with your family and friends.

Further challenges to dominant meanings and institutionalized norms appear in articles ranging from the plight of homeless teenagers to contemporary political discourse. In the case of the latter, a young woman shares her observations about the 2000 Democratic National Party Convention:

> My liberal, idealist heart retreated into a sort of superior cynicism, inwardly sneering at the people on the convention floor who so enthusiastically accepted whatever their centrist, poll-driven leaders presented as the Ideal of the Month. I thought about voting for Ralph Nader. I thought about not voting. Then I thought about what I believe in. I do believe in reproductive choice, gun control, affirmative action, raising the minimum wage, and campaign finance reform. I want to fund higher education, Social Security, Medicaid, Medicare, Head

Start, and other social services. I think it's the government's job to ensure that businesses aren't polluting the environment and the every citizen has health care. Ignoring the commercials, the speeches, and even the candidates, I suddenly realized that elections are important. Parties are important. Politicians have real power over questions that affect our lives. . . . This article is starting to sound a bit like some of those speeches I heard at the convention, but perhaps that means that the people who spoke those words meant them as sincerely as I do these.

An article critical of media monopolies contained useful information about and links to the media watchdog group Fairness and Accuracy in Reporting (FAIR):

> As media conglomerates and media bias become more and more prominent, public resistance grows. . . . FAIR works with both activists and journalists by exposing important news stories that are neglected, advocating for greater diversity in the press and providing constructive and well-documented criticism of censorship and media bias.

Affirmative action was addressed in an essay concerning college preference policies that favor applicants from certain racial categories. "Being of Chinese descent, I am part of the 'over-represented minority' that, along with Caucasians, is allegedly hurt by affirmative action. . . . But the truth is, I benefited from affirmative action. And so does every single college applicant today." She then argues that increased levels of diversity and the positive aspects of interacting with and learning from people from different geographic, economic, or racial backgrounds outweigh any "flaws" in the admissions process.

These articles not only serve a practical function by providing girls with opportunities for taking further action, but they also serve an ideological one by introducing readers to language and symbols relating to contemporary political and social issues, further reinforcing the relationship between activism and social change.

Encouragement for readers to question educational norms appeared in two articles. In one, a young woman describes her feelings stemming from the need to be perfect in college. She cautions other "over-achieving readers" to excel at their strengths but to also strive for a more well-rounded experience. "I wish I could take that frantic college girl who had suffocated from perfection and scream, 'Go out and have fun. Another perfect grade is not worth it. Make friends, do an internship in a field you enjoy, and experience the world outside the school's walls.'" In a separate essay titled, "The Feminist Sorority Girl: Not a Contradiction," a young woman defends the Greek system's potential benefits for female students.

> Serious people with serious goals can also enjoy this social environment because it enables them to form lifelong friendships and establish a connection to a national community of strong women. Those who participate in social justice organizations, student government, literary magazines and athletic teams in addition to sororities do not need to segment themselves into "sorority girl" and "serious person." Not only can these two components coexist, but they often promote each other. . . . As feminists, we should welcome organizations

coordinated by and made up of women, and recognize our own ability to empower and influence the Greek community and our sisters.

While acknowledging that all sororities "would do well to ask themselves how they can be more welcoming of potential sisters of varied ethnic backgrounds so that diversity can be," she concludes that there have been a growing number of sorority women in the U.S. who are occupying leadership positions and/or participating in programs such as Model United Nations and Take Back the Night.

Girls expressed power and agency in articles about technology, science and the Internet, and in features such as "Hear from the Pocket," a forum where girls speak out about current events and/or issues,[10] and "Ask Dr. Beth," an advice column that tackles weighty issues like depression, drugs and alcohol, and self-mutilation.[11] Feminist perspectives were also found in book reviews for *Shattered: Stories of War and Children*, "a juxtaposition of youth and war" and *It's About Time*, "a book by and for women . . . [about] relationships, rights, futures, bodies, minds, and souls." Similarly, a movie review of *The Hours* described it as "a devastating portrait of three women fighting for inner happiness . . . [that] asks the viewer to consider what constitutes oppression and what can lead to true liberation."

Conclusion

The majority of Blue Jean Online content analyzed explored the intersections between gender, power, and social interactions, making feminism visible and accessible to girls. Clearly, the ability of Blue Jean Online to position itself as a feminist alternative to mainstream girls' magazines is largely due to an environment in which the economic necessity of advertising does not determine the production of content. Indeed, unlike *Seventeen* and *Cosmo Girl*—which bombard readers with multiple pages of advertisements before featuring a single article—visitors to the BJO Web site are only occasionally met with pop-up ads for *Ms.* magazine or Handel's book. This finding is consistent with those from other studies that have demonstrated that the presence or absence of pressure from advertisers to maintain the status quo directly impacts editorial decisions. What remains unclear is whether Handel can continue this practice given the economic uncertainties associated with operating a non-profit organization.

Of equal importance is the way in which meanings about gender at the production, text and reception levels provide girls with opportunities to construct their own feminist identities. Specifically, Blue Jean Online offers girls rhetorical spaces where dominant meanings relating to gender roles in societies can be negotiated, resisted, or rejected outright. Girls and young women not only occupy a variety of subjective identities within the magazine—accompanied by varying degrees of power—but also self-identify as feminists, suggesting that negotiations over the construction of meanings and values about femininity do occur. On a more practical level, Blue Jean Online calls on girls to act, providing them access to outside resources and materials that support feminist causes and viewpoints—very often linking them to Web sites that promote social and political activism. In so doing, the magazine encourages girls to become active seekers, rather than passive receivers, of information.

Blue Jean Online is not an ideal magazine for girls. Signs of commercialism do appear from time to time, as when book reviews provide links to Amazon.com's Website. Additionally, while a wide range of racially and ethnically diverse viewpoints were expressed by girls within the U.S., more effort needs to be placed on gaining international input, especially since the magazine's mission statement emphasizes an international focus. Further audience research is also needed before assumptions can be made about whether BJO and similar girl-zines appeal to girls who are not already inclined to prefer alternative meanings about femininity.

As a feminist researcher, I am mindful of the danger in assuming that ideological change is occurring in girl culture based on an analysis of one Internet magazine. Indeed, the media are known to frequently intersperse small amounts of a liberal feminist discourse in with the dominant discourse to accommodate their critics. At the same time, in a postfeminist environment in which mainstream media tend to treat feminism as outdated and irrelevant to females' lives, we must continue to make clear distinctions between media that advance feminist causes and media that co-opt a feminist subtext. Add to that Steiner's challenge, and Blue Jean Online emerges as a viable feminist alternative to traditional girls' magazines that *stands up for girls and young women.* Whether this represents any sort of hegemonic shift in popular culture remains to be seen.

Notes

1. For example, Budgeon and Currie (1995) have described feminist content as that which appears to be "potentially political" in nature because it draws attention to social problems or political issues, questions the appropriateness of the status quo, and/or urges readers to take some form of political action.
2. The message about "responsibility" has many manifestations. In *Seventeen* and *YM*, for example, girls are taught that a successful prom night depends on their ability to balance controlling the way their dates dress and behave with making sure they have a good time (Mazzarella, 1999).
3. A cursory study I performed of eight girl-zines revealed that several challenged the dominant discourse (Walsh, 2002). For example, Web sites for *Smart Girl*, *Teen Voices*, and *Terrifichick* conveyed feminist messages regarding a number of issues.

4. According to the BJO Web site, Sherry Handel, CEO and President of New Generation Media, Inc. began publishing *Blue Jean Magazine* in 1996 as an alternative to traditional beauty magazines. She ceased printing the publication in 1998, claiming at the time to have an international circulation of 300,000, and in 2000 created Blue Jean Online. It appears that, initially, she accepted some advertising. However, in 2003, apparently frustrated by her lack of control over online marketers, she cancelled a contract with a third-party advertising firm. It is also interesting to note that up until 2003, Handel had maintained that she planned to resume publishing the print version of the magazine, then eventually reconsidered that plan due to "financial constraints." Clearly, her decisions were the result of inherent tensions that have existed historically between editorial and financial decision makers and that have subsequently forced *Ms.* and other politically and socially progressive magazines to reject revenue from advertising.

5. A page titled "Our Partners" indicates that Blue Jean Online also has connections to feminist resources like WomensRadio.com, NationalWomensCalendar.org, and Women's Express, Inc.

6. Content was closely examined at three separate intervals throughout the twelve-month period: October/November, 2002; July/August, 2003; and October/November, 2003.

7. The story contains links to related Web sites including Kilbourne's.

8. Well-known women such as The Body Shop's founder Anita Roddick, cartoonist Cathy Guisewite, columnist Molly Ivins, media researcher Jean Kilbourne, and former Secretary of State Madeline Albright were among those featured—as were newcomers like Victoria Nam, author of *Yell-Oh Girls*, an anthology about growing up as an Asian American girl. (See the Blue Jean Online archives for the complete list.)

9. *Vagina Monologues* is a play based on Ensler's interviews with 200 women about domestic violence, rape, female genital mutilation, and other forms of abuse. The adults who complained were a parent and a Catholic school principal.

10. For example, readers were asked to respond to the following questions: What do you think about "Operation Iraqi Freedom" led by American and British forces—Do you think that the Iraqi people want to be freed from Saddam Hussein and his regime, or is the invasion unwanted by the Iraqi people?

11. It is worth noting here that Dr. Beth Jelsma, a psychotherapist with a Ph.D. and twenty-years experience working with teens has a good deal more credibility than the editor-in-chief of seventeen.com, who also dispenses advice to girls about dating, friendship, drugs and sexuality.

References

Budgeon, S. (1994). Fashion magazine advertising: Construction of femininity in the "postfeminist" era. In L. Manca &, A. Manca (Eds.), *Gender and utopia in advertising* (pp. 62–70). Lisle, IL: Procopian Press.

Budgeon, S., & Currie, D. (1995). From feminism to postfeminism: Women's liberation in fashion magazines. *Women's Studies International Forum, 18*, 173–186.

Carpenter, L. (1998). From girls into women: Scripts for sexuality and romance in *Seventeen* magazine, 1974–1994. *Journal of Sex Research, 35*, 158–169.

Center for Media Education. (2001). *Teensites.com: A field guide to the new digital landscape.* Retrieved December 12, 2001, from http://www.cme.org/teen/study

Chonin, N. (1999). Girl wide web. In K. Massey (Ed.), *Readings in mass communication* (pp. 84–88). Mountainview, CA: Mayfield Press.

Currie, D. (1999). *Girl talk: Adolescent magazines and their readers.* Toronto: University of Toronto Press.

Duffy, M., & Gotcher, M. (1996). Crucial advice on how to get the guy: The rhetorical vision of power and seduction in the teen magazine *YM*. *Journal of Communication Inquiry, 20*, 32–48.

Duke, L., & Kreshel, P. (1998). Negotiating femininity: Girls in early adolescence read teen magazines. *Journal of Communication, 22*, 48–71.

Durham, M. G. (1998). Dilemmas of desire: Representations of adolescent sexuality in two teen magazines. *Youth & Society, 29*, 369–389.

Fiske, J. (1989). *Understanding popular culture*. London: Unwin Hyman.

Garner, A., Sterk, G., & Adams, S. (1998). Narrative analysis of sexual etiquette in teenage magazines. *Journal of Communication, 48*, 59–78.

Goldman, R. (1992). *Reading ads socially*. New York: Routledge.

Gramsci, A. (1971). The study of philosophy. In Q. Hoare & G. Nowell-Smith (Ed. & Trans.) *Selections from the prison notebooks* (pp. 323–380). New York: International Publishers.

Hall, S. (1980). Encoding/decoding. In D. Hobson, A. Lowe, & P. Willis (Eds.), *Culture, media and language: Working papers* (pp. 128–38). London: Hutchinson.

Harding, S. (1991). *Whose science? Whose knowledge?* Ithaca, NY: Cornell University Press.

Jaggar, A.(1988). *Feminist politics and human nature*. Lanham, MD: Rowman & Littlefield.

Jansen, S. (2002). *Critical communication theory*. Lanham, MD: Rowman & Littlefield.

Kearney, M.C. (1998). Producing girls. In S. A. Inness (Ed.) *Delinquents & debutantes: Twentieth-century American girls' cultures* (pp. 285–310). New York: New York University Press.

Lamb, L. (1999). Media criticism: The unchanging nature of girl's magazines. *New Moon Network, 7*, 14–15.

Markham, K. (1993). Seventeen *magazine's construction of femininity: A socially informed semiotic analysis*. Unpublished doctoral dissertation, University of Oregon, Eugene.

Mazzarella, S. R. (1999). The "Superbowl of all dates": Teenage girl magazines and the commodification of the perfect prom. In S. R. Mazzarella, & N. O. Pecora (Eds.), *Growing up girls: Popular culture and the construction of identity* (pp. 97–112). New York: Peter Lang.

Merskin, D., & Silva, K. (2000, November). Trauma-Rama: Confession columns in *Seventeen* magazine. Paper presented at the meeting of the National Communication Association, Seattle, WA.

Peirce, K. (1990). A feminist theoretical perspective on the socialization of teenage girls through *Seventeen* magazine. *Sex Roles 33*, 491–500.

Rowe, C. (1997, January). *The book of zines: Readings from the fringe*. Retrieved on November 16, 2001, from http://www.zinebook.com/whatcha.html

Steeves, H. L. (1987). Feminist theories and media studies. *Critical Studies in Mass Communication, 4*, 95–135.

Steinem, G. (1990, July/August). Sex, lies and advertising. *Ms*, pp. 19–28.

Steiner, L. (1995). Would the real women's magazine please stand up . . . for women. In C. Lont (Ed.), *Women and media* (pp. 99–108). Belmont, CA: Wadsworth.

Stern, S. (2002). Virtually speaking: Girls' self-disclosure on the WWW. *Women's Studies in Communication, 25*, 223–253.

van Zoonen, L. (1994). *Feminist media studies*. Thousand Oaks, CA: Sage.

Walsh, S. (2002, November). The 'scoop' on girl-zines: Alternatives worth considering. Paper presented at the meeting of the National Communication Association, New Orleans, LA.

5

Exploring Dora

Re-embodied Latinidad on the Web

Susan J. Harewood
and Angharad N. Valdivia

Introduction

The word "web" has been a particularly evocative metaphor for our understandings of new technologies. By evoking both the gossamer reach of spider webs and the danger of the spider lurking at the center, the word seems simultaneously to signal both our utopic and dystopic visions of cyberspace in a gendered manner. After all Charlotte was a female spider. Nevertheless, though this metaphor has been suggestive of a great deal, we will argue in this chapter that, thus far, academic and popular imaginings of the web of cyberspace have been rather truncated. In much academic work cyberspace is seemingly thought of as a "world apart," as if the Web does not extend much beyond the computer screen, as if it does not reach out and articulate with other common webs of communication and representation in the everyday. In this chapter we will suggest that the web metaphor can be usefully extended in order to recognize the interrelations between representation in cyberspace, representation in other forms of popular culture, and representations of the body and the self. We will argue that seeing the Internet within the web of popular culture and social relations prompts us to consider the disembodied subjectivity that the Internet allows in relation to the embodied and material experiences of Web users. Specifically we argue that this disembodied subjectivity must be considered in relation to the embodied and material experience of contemporary gender and ethnic politics in the United States, a struggle which, given recent demographic and marketing developments, inevitably includes Latina/os and Latinidad.

To explore and navigate this web of metaphors, representations, and identities we focus on one particular character, Dora, the star of the television show *Dora the Explorer*. We choose this show because it allows us to explore the negotiation and struggle over this character. These struggles are indicative of pedagogical strategies as well as processes of ethnic differentiation in the contemporary United States. Furthermore, we choose Dora because she is so easily accessible in the mainstream—available many times daily on both cable packages through Nick Jr. and on over-the-air television stations through CBS. Whereas those living in cosmopolitan settings may have access to a broad range of cultural forms and those identifying with a particular ethnic or cultural group may seek out particular types of media programming, a show such as *Dora* can easily be reached by anyone owning a television—something more or less universal in this country.

We will be specifically conducting a multi-sited examination of representations of and negotiations and struggles about *Dora the Explorer* as the Nick Jr. *Dora* Web site and message board participants interrelate with the *Dora* television program, with the current national tour of *Dora Live!* and the popular press coverage and marketing of Dora in all of her many forms and locations. Given that the Web site seems to be particularly linked to the traveling *Dora Live!* show, we investigate the struggle and negotiation over Dora's language and body as represented in these three cultural sites. We argue that, despite the rhetoric of "disembodiedness" that often accompanies the Web, its representations, and its participants, the body follows the narrative, repeatedly reinserting itself as a way of enforcing and policing borderlines about ethnicity and mainstream culture. Dora reminds us of the impossibility of leaving the body behind in any kind of form of popular culture because people are always bringing the body back into discussion and embodying the representational, which itself embodies dominant tropes of an ethnicity. Dora signs in as a Latina in a representational terrain that continues to foreground whiteness and background other ethnicities in relation to whiteness. A major trope of ethnicity, especially concerning Latina/os, is language—that is, Spanish. In the *Dora* situation, language becomes another major issue of struggle and a conveyor of discordance over Dora's body. Through language, which is absolutely an essential component of Web chat, the body is reintroduced and with it ethnicity and difference. We will argue that both Nick Jr. and many message board participants construct Dora's Latina identity by drawing from a multiculturalist meaning system that views Latinidad as a resource from which one can derive instrumental educational gains and employment opportunities while simultaneously maintaining the centrality of whiteness and othering and exoticizing Latino/as.

The Literature

Literature on new technologies has been marked by a number of telling absences that have tended to identify the "cybercitizen" as white, male, and adult. This is clearly part of the discourse of modernity. Much of this discourse—a discourse that nearly always ignores the historical context of development and implementation of "new" technologies (e.g., Marvin, 1988)—foregrounds the use of the Internet and the Web as if these were new issues with unlimited emancipatory potential in a post-gender and post-racial world. This framework practically predetermines a situation in which issues of gender and ethnicity are absent from most popular culture discussions. For example, the modern subject, the one who engages with the latest technology, lifestyles, and mobility vehicles has long been configured as the white, EuroAmerican, male subject. Feminist theory has been valuable in documenting the exclusion of women from these discourses of modernity (e.g., Felski, 1995), however, gender studies has also highlighted that when women *are* taken into consideration there has been a propensity to focus on the Western, white woman (Mohanty, 2003). Thus when we examine cyberfeminist literature we recognize that, though it is marked by a commitment to the emancipatory promise of cyber technologies, insufficient attention has been paid to issues of ethnicity (McLaine, 2003; Sangha, 2004; Suden, 2001). As Kolko, Nakamura, and Rodman (2000) and Nakamura (2002) attest, despite the complicated matrix of identity proposed by Donna Haraway (1991) in the immensely influential "A Manifesto for Cyborgs," most of the cyberfeminist literature has largely ignored racial components in that essay when they apply it to an analysis of the Internet. Moreover, when race is addressed, it is addressed through a binary racial economy of white and Asian. In new technology literature Asians and Asian Americans are recast as the "other" in relation to whiteness, thereby supplanting the more usual black other. Reasons for this revised racial dichotomy, which nonetheless maintains whiteness as the central location and ultimate referent, are discussed elsewhere (Nakamura, 2002). However the result is that Latina/os fall out of this digital landscape altogether, either as referents for the digital or for the real. In fact, studies on the communications divide, a divide that exists and is usually widened with the introduction of each new telecommunications technology, suggest that the digital divide for "Hispanic" Americans continues to increase even if one includes income and/or education variables in the study (Hacker and Steiner, 2002). Whereas Internet usage has increased among all ethnic categories, differences in usage remain with Hispanics trailing blacks, whites, Asian Americans, and Pacific Islanders (U.S. Department of Commerce, 2002). While there is little research conducted on Latina girls or adolescents and their use and access to the Internet, a recent study by Mayer (2003) documents that in a working-class community, where economic resources are scarce, the adolescent girls she studied had very little access to media technology at home. Even television and CD-player technologies were not to be assumed. In this group and community she found very little evidence of computer technology.

An additional exclusion in issues of new technology applies to age. The discourse of modernity and technology also leaves out the young and old, foregrounding the adult middle-aged male. When scholars *do* address age as a factor, as within the long tradition of research on children and the media, they tend to use children as a disaggregated group. For instance, Livingstone and Bovill (2001), in an edited collection that focuses on children, follow this disaggregated strategy in all but one of fourteen chapters. Even though within gender and feminist studies there is a growing body of work on both girls and adolescent females, the original impetus in new technology scholarship tends to focus on boys. Lemish, Liebes, and Seidmann (2001), in their cross-national literature review find that "On the whole boys' bedrooms are more high tech than girls' bedrooms . . . [and] boys are attracted to 'new' media, whereas the familiar 'old' media have greater following among girls." (pp. 266–267). Given that Internet usage, however, is higher among children than young adults (U.S. Department of Commerce, 2002), one would expect that boys would dominate this new technology. However, this is not necessarily the case. Whereas early reports of Internet use among females found the highest level of use among girls between the ages of twelve and seventeen, as a whole, research showed that by the year 2000 women's presence online surpassed that of men (Rickert & Sacharow, 2000). More recent investigations have seen that gender gap disappear in that usage tends to have equalized among the genders (U.S. Department of Commerce, 2002). Thus, although boys have been found to have more of an orientation towards new technologies, there is no gender gap in terms of usage of this technology. Whether due to gendered patterns of sociability, gendered patterns of family relationships and media use, or gendered social values and self-image, Lemish, Liebes, and Seidmann (2001) conclude that boys and girls use media differently, yet they both use the media, both the old and the new media, even though they largely prefer drastically different types of content and narratives.

Usage and deployment of the Internet follow many of the same trajectories as other forms of communication technologies. Despite their original interactive potential, they end up as commercial, centralized transmitters with many receivers/consumers who are, at most, engaged in a reactive form with the medium. Given that so much of Internet research and discourse focuses on the potentially interactive and identity-forming activities enabled by the Web, contemporary research explores the sites where this may be occurring. This shifts the focus from consumption of Web messages to production of identities and spaces of creativity or relative narrative freedom. The Internet, despite its over-foregrounding of commercial messages, may still contain a space for interactive participation. Foremost among these are blogs—spaces wherein participants can carry on a conversation or narrative about their chosen topics. Research on blogs and girls, usually focusing on the concept of self-disclosure, suggests that girls are indeed very active in this area (Stern, 2002) and that their chat includes body talk, sometimes exclusively as in Pro-Ana sites (Mastronardi, in press). Thus, though blogs represent an overlap of journal writing or bulletin board posting with the new Internet technology, in fact, as Ellen Seiter (2002) reminds us, "as the Internet develops from a research-oriented tool of elites to a commercial mass medium, resemblances between Websites and television programming increase" (p. 35).

This is precisely the point at which Dora exists. Her television show represents the intersection between television and computer imagery as the whole show is like playing a marginally reactive video game. Simultaneously the chat rooms on the *Dora* Web site present not so much viewers but their parents, guardians, and older siblings with the opportunity to participate in a conversation about the television show, the live show, any products, and indeed the entire *Dora* phenomenon. As part of this larger web of popular media consumption the Internet is open to the scrutiny of popular culture research. Three intersecting areas that are particularly relevant as we examine Dora are research that examines the intersection between media and Latina/o Studies, popular culture and pedagogy research that focuses on education and multiculturalism, and popular culture and pedagogy research that highlights the influence of popular culture in students' lives.

Research on the intersection between Latina/o and popular culture studies is huge and varied (see for example *Communication Review* [Latina/o communication, 2004]). Issues of relevance to this project include efforts to forge a pan-Latinidad that acts as an umbrella for a number of singly or multiply hyphenated identities such as Mexican American, Puerto Rican American, or AfroCuban Latino. As Mayer (2004) discusses, we still need to work out many of the theoretical complexities of this newly created umbrella term, a term that comes both from within Latina/o communities as well as from governmental efforts to account for and categorize the population according to ethno-racial categories. As a result of this effort to construct this new Latina/o audience (Dávila, 2001), it makes much more marketing sense to construct a mestizo audience of ambiguous specificity (see also Halter, 2000) than to target much smaller individual communities especially given the undeniable demographic growth of the Latina/o population, which as of February 2003 was announced to be the largest minority group in the U.S. Nonetheless, there are competing discourses of Latinidad within the mainstream.

The traditional Mexican American paradigm is much more nation specific, Latin American referent, Spanish dominant, patriarchal family style, and working class than the modern synchretic model which proposes an upwardly mobile, usually single, economically successful, U.S. Latina/o referent, pan-Latina/o identity (Dávila, 2001; Levine, 2001). This synchretic option while new, nonetheless, draws on tropicalizing tropes (Aparicio and Chávez-Silverman, 1997), which assign "traits, images, and values" (p. 8) to anyone south of the U.S. border. While many of these traits, images, and values refer to adults, especially women, some of these tropical overtones of island iconography, palm tree-lined settings, with coconuts and monkeys, and a soundtrack of tropical music—namely, salsa or merengue—can be deployed on a children's show, and they are deployed on *Dora*.

Education and multiculturalism research have highlighted the conservative uses of multiculturalism discourses in order to maintain the centrality of the traditional white Western canon (Appadurai, 1996; Giroux, 1991; McCarthy, 1998; Mahalingam & McCarthy, 2000). *Dora* reinforces these findings but also flips them over in that *Dora* asks us to examine the ways in which educational styles and issues become part of children's popular cultural products. Thus we seek to ask, not only how the classroom experience is shaped by popular culture, but how children's popular culture is embedded within the discourses of education, including specifically those mainstream discourses that protect white privilege. In other words because the program is crafted by an educator, those discourses in turn feed back into children's programming.

Exploring Dora

This chapter provides a critical cultural analysis of inter-related and inter-textual locations of *Dora the Explorer*. Against the backdrop of Dora, the character, the essay focuses primarily on an analysis of chat streams accessible through the official Nick Jr. Dora the Explorer Web site, wherein through the "Dora for parents" portal one can enter the message boards. In these message boards we selected the chat streams that highlighted major issues of identity as identified in contemporary Latina/o studies literature: the body, language, and territorial/national origin. All three of these issues, as documented in the above literature review, are major signifiers of a contested Latinidad within mainstream popular culture. Over the period of a month we monitored the Nick Jr. Web board, identifying fifteen threads that addressed these issues. Five of those threads maintained a relatively extended conversation allowing for an examination of the intersections of Web practices, biographical identities and broader meaning systems (Denzin, 1999). We will offer intensive analysis of these five key threads.

There are many reasons why *Dora* presents us with an ideal case study. Firstly, *Dora* appears in so many different media. She was initially introduced in 2000 on the Nick Jr. Web site, and this was swiftly followed by the premiere of the television program which takes its stylistics from CD-ROM technology (complete with cursor and pop-up windows). The popularity of *Dora* has spawned a wealth of synergistic merchandise that has branched out beyond the products one might expect—such as CD-ROMs, computer games and videos—to include a range of everyday merchandise including cards, clothing, furniture, food, beauty products, wallpaper, stationery, school and party supplies, and so on, culminating in the national tour of the live show, *Dora Live!* Secondly, *Dora's* ubiquity seems to stand in such stark contrast to the absences outlined above: the absence of women, the absence of girls, and especially the absence of Latinas not only from popular media including new media but also from academic thinking about those media. In fact *Dora's* unique position has formed an important part of the discourses surrounding her character. Both on the official Web site and in the popular press, the little animated *Explorer* is frequently identified as some sort of pioneer on a journey towards Latinidad's acceptance in a North American mainstream culture. When asked on the Web site, "What's different about Dora?," one of the creators, Chris Gifford, highlights *Dora's* differences:

> Probably the most obvious is the fact that we teach Spanish vocabulary in every show and that Dora is a Latina. Another unique aspect of the show is that it stars a little girl as an adventurer. The way we incorporate our curriculum into the show is different. (Gifford, 2004)

Members of the popular press have suggested that the creation of *Dora* is part of the growing representation of "multicultural themes" on children's television in general (Barney, 2002; Bauder, 2003) and as a challenge to negative Latino/a stereotypes in particular (Martinez, 2002). The *New York Times* reported Nickelodeon officials as identifying *Dora* as part of their push to "reflect the world children lived in" (Navvaro, 2001). It is a world in which, as was mentioned earlier, Latina/os are the most numerous minority in the country. Thus, given that it is the market that drives much of programming in a capitalist culture, it makes sense to appeal, by way of representation and [virtual] inclusion, to this growing segment of the population. Latinos under thirty-five have $300 billion in annual purchasing power (Bauder, 2003). However, *Dora's* main audience is white, and a careful re-reading of diversity discourses about *Dora* demonstrate that the push towards "good stories and *universal* story lines" (Navvaro, 2001 emphasis added) reveals an additive notion of diversity. Such a notion of "diversity" and "multiculturalism," as we suggested above, has been particularly explored in curriculum studies (Appadurai, 1996; Giroux, 1991; Mahalingam & McCarthy, 2000; McCarthy, 1998). And it is a literature that is relevant with children's programming in general and *Dora the*

Explorer in particular because of the expressly pedagogical objectives of this genre of popular culture. *Dora the Explorer* is targeted at preschoolers and, according to NickJr.com, the program has specifically educational objectives:

> *Dora the Explorer* teaches children how to observe situations and solve problems as they explore Dora's world with her. Along the way, kids learn basic Spanish words and phrases, as well as math skills, music, and physical coordination. The show is highly interactive,[1] and Dora's young viewers are encouraged throughout the show to respond to Dora and to actively participate in the adventure through physical movement. (NickJr.com, 2004)

These educational objectives can provide us with a framework for perceiving the ways *Dora* exists as part of the large web of popular culture. By understanding *Dora* as part of popular culture, we can then investigate the ways in which Internet experiences remain grounded in our everyday lived biographies (Denzin, 1999).

Whereas Dora's Web site includes educational objectives and the desire to contribute to liberal multicultural discourses, there is nothing to be found about any type of Latina/o Studies basis for the television or live show and the Web site. In other words, there is no indication that this character was constructed with input from any scholarship that considers issues of not only children and the media but also of ethnic studies and issues of representation. The implicit understanding that the intended audience is a mainstream audience—that is, mostly white and middle class—adds to the show's additive multiculturalism approach, one that builds on traditional tropes of ethnicity in general and about Latinidad in particular.

Dora and the importance of tropical place

On the *Dora* television program, observation and problem solving learning objectives revolve mostly around the need for Dora to reach a particular destination so that she can help someone in distress. For example, in Episode 114 entitled "Sticky Tape," Dora, Boots (her constant companion who is a monkey who wears large red boots), and the audience need to get sticky tape to Benny the Bull to patch his hot air balloon before it sinks into Crocodile Lake. A recurring character called The Map plots out the path to Crocodile Lake, which requires Dora and Boots to cross The River, and go over The Rock until they reach Crocodile Lake. In each program children are expected to remember the three locations that get Boots and Dora to their objective, and they are also expected to respond to Dora who asks them at each point where she is to go. Dora waits, by general television standards, quite a long time for the children to identify the location on the screen and shout the answer at the television. However, it must be noted that this is a reactive relationship as Dora's part is pre-written, regardless of what the children are doing in their individual viewing situations. *Dora* and its producers do not hear what the children are saying. This, as we will later see, is also the case on the Internet, as those posting in the chat rooms often identify themselves as parents or older siblings.

Nonetheless, the emphasis on location in problem solving is understandable for a program that centers on an explorer. However, meanings made of those locations by *Dora* parents on the message board and by Nickelodeon demonstrate broader issues of location and representations of ethnicity. In fact, it is only through the Internet that parents, guardians, older siblings, and assorted other *Dora* watchers can interact with each other in their discussions about the show. In the message board thread "Where is Dora????" *Crossers*[2] asks:

> Where is Dora from? We have been trying to identify the region based on background scenery, the animals and, of course, the dialect. Jungles, mountains, palm trees, monkeys, lizards and snails. Where or where can Dora be? (NickJr.com, 2004)

Most of the answers to this query seek to emphasize that *Dora's* environment is make believe and should not be taken seriously. For example *hmicaela*, a regular respondent on the message board answers:

> Just south of Nowhere Land, East of Never Never Land, West of OZ and right next to the land where the Dragon Tales are. J/K

> Seriously, I don't think it represents any particular region, but a big latin-inspired fantasy-land. (NickJr.com, 2004)

horamut, another frequent visitor to the board, answers, "Dora lives in a virtual Latin American country inside a computer. End of discussion. :)" (NickJr.com, 2004). *MAMASONA* emphasizes Dora's pan-Latina "heritage": "She is Latina, the creators have never given a name to her home." (NickJr.com, 2004). This post refers to the unspecific pan-Latina identity promoted by the program and Web site information wherein the program and the character are described as follows: "*Dora the Explorer* is a half-hour animated children's television series starring a 7-year-old Latina girl and her friends" (NickJr.com, 2004). Postings on the Web board seem to indicate that those participating in the exchange perceive of the *Dora* environment in at least four ways. Firstly, they seem to think that the 2D environment of "Jungles, mountains, palm trees, monkeys, lizards and snails" references Latin America in some way. Secondly, they do not seem to think the specificities of a particular Latin American location and nationality are particularly important. Thirdly, they appear to have internalized tropicalizing tropes, equating all of Latin America within the geocultural trope of the tropics. Fourth, they reiterate, with the Web site itself, that the dominant culture in the U.S. views Latina/os as eternal outsiders.

This is also true in the thread "What is COQUI????????". In this thread, parents continue to negotiate Dora's nationality. *Pattysanch* asks:

> On the show, Dora, when she talks about a frog, she calls it COQUI. I thought frog in
> Spanish meant SAPO or RANA. The word coqui does not exist in the Spanish dictionary.
> I have heard it before, when I was in Puerto Rico once, and it was the name of a frog they
> have there. So does that mean Dora is Puerto Rican??? (NickJr.com, 2004)

hmicaela, *horamut* and *MAMASONA* respond again. *hmicaela* suggests that the
name Coqui references a type of frog or toad, and adds "I don't think they specify
Dora's nationality" (NickJr.com, 2004). *horamut* suggests the name Coqui references
a Latin American folk tale. *MAMASONA* reiterates the pan-Latinidad, "As I have
quoted before the creators have left it blank so she represents all Latin
cultures."(NickJr.com, 2004)

However, other respondents are less inclined to leave Dora's nationality so
open. *pinkdaze* states emphatically that Dora is Puerto Rican and *FunMomma04*
believes that Dora was once referred to as "the Little Mexicanita." Though *pinkdaze*
and *FunMomma04* are quite emphatic about Dora's nationality, it still seems that
location as an indicator of Latinidad is relatively open. In other words, general "Latin
Americaness" is more important than national specificity; and "Latin Americaness"
should be understood and emphasized as tropical. In this, therefore, we are seeing
the ways in which *Dora*, whether as television program, Web site or Web board, con-
forms to the long history of hegemonic tropicalizations (Aparicio & Chávez-
Silverman, 1997) found in representations of Latin America and U.S. Latina/o
identities and cultures. Linking tropicalization to orientalism, Aparicio and Chávez-
Silverman suggest that tropicalization seeks to construct Latinidad as the ever-
foreign, exotic Other. One part of this trope is the privileging of "the taxonomy of
flora and fauna over the recognition of the land's inhabitants" (Aparicio & Chávez-
Silverman, 1997, pp. 9–10).

The popular media's reinscription of this tropicalization is particularly apparent in the description of the setting of *Dora Live!*: "The interactive show invites audience members to help Dora solve puzzles and complete her journey through a
colorful tropical world of beaches and rainforests" (*This week in Chicago*, 2004).

The delineation and non-delineation of *Dora*'s environment bring to the fore
two themes that recur across the message board and seem important to the Web
board's contributors' meaning-making processes. These are: What signifies Latina/o
cultural identity? To what extent do the virtual specificities of *Dora the Explorer* matter "in real life"? A particularly unreflective comment in the *New York Times* highlights the ways in which Latina/o cultural identity is signaled in popular culture in
very narrow ways:

> The creators, Chris Gifford, Eric Weiner and Valerie Walsh, gave Dora no distinct nation-
> ality, so that more viewers identify with her, but she has a Hispanic look, uses Spanish words
> (like "vamonos, let's go") and has adventures with a decidedly pan-Latin flavor. (Navvaro,
> 2001)

It would appear that the particularities of Latinidad can be substituted for vague tropicalization in the name of ratings. Thus the Latina/o is identified, therefore, as being ever the tropical outsider who can be known by a specific "Hispanic look" and by his/her use of Spanish. Similarly those posting on the Web board seem to be generally willing to accept the vague tropicalization as sufficiently representative of Latina/o cultural identity, but they are less flexible about the look of the embodied Dora in *Dora Live!* And they are particularly meticulous when it comes to Dora's use of Spanish. Thus Web board participants seek to construct a Latina/o authenticity in specificities about the body and about language.

Dora and the authentic body

The discourse of authenticity is often deployed as a way to draw boundaries around identity categories. Those within are authentic and those outside are not. Within some ethnicity approaches, especially those bordering on biological definitions of race, a major way to draw boundaries is through the body. Thus, although the chat rooms on the Web ostensibly represent disembodied discussions, the body reinserts itself in two very similar threads, "Dora Live Show-ethnicity?" and "Dora Live: 'Who's the White Skinny Girl??'" It becomes clear that for many parents the embodied Dora should be as brown and as chubby as they interpret the animated Dora to be. For a number of the Web board participants, Dora's brownness is representative of her authentic Latina identity. For example, *goldengram* who begins the "Dora Live Show-ethnicity?" thread says:

> I have been waiting for Dora to come to town, but seeing the commercial has made me weary. It looks as though a Caucasian girl is playing the role of Dora. If that is the case, I have a real problem with that. I am sure that there are plenty of young Hispanic or Latina girls who are talented enough to play the role of Dora. (NickJr.com, 2004)

Some of the parents who see Dora's skin color as being important tie it to their own embodied experiences. Both *goldengram* and *melvin01* identify themselves as Black. *mebesand* who is also particularly concerned with *Dora Live's!* slim build (*mebesand* started the "Dora Live: 'Who's the White Skinny Girl??'" thread) identifies her daughter as Mexican American. These parents, and others who do not make either their own or their children's ethnicity apparent, suggest that their children will not be able to relate to the fair-skinned actress who plays Dora. Their responses highlight our contention in this chapter that, despite the emphasis on disembodiedness in new technologies literature, it is impossible to leave the body behind. Of course, as much of the literature suggests, there are no guarantees that the people who state their ethnicity are, in fact, members of that particular group in real life. Nevertheless, the Web board participants return again and again to the body—theirs, their children's and Dora's. And the meanings of the ethnic body, as they are deployed by these Web board participants, continue to be enmeshed within the webs of the dominant tropes of representations. The Web board participants provide us with relatively clear examples of the ways in which we come to think through the politics of the body. The parents above who are concerned that *Dora Live!* is neither brown nor chubby enough seem caught within one of the dilemmas of political/the politics of representation. That is to say, political representation has traditionally required stable categories from which to launch a challenge to oppressive systems, yet as these categories are often delimited by the very structures of those oppressive systems, emancipatory efforts can become self-defeating (see Butler, 1990). Thus, in this case on the *Dora* Web board, parents who have consistently confronted their own erasure from popular culture, in demanding political visibility, end up using categories that set up rigid, essentializing boundaries around Latinidad. The deployment of strategic essentialism for political and cultural purposes, something that has been discussed by Fusco (1995), appears in these chat rooms when parents and others make representational demands from the traveling show. If we are to have a Latina character, for whatever reasons, these posters seem to say, then she must abide by stereotypical compositions of the Latina body so that she remains recognizable as a Latina. In other words, in seeking representation, ethnic tropes are reiterated and solidified. Dora must remain brown and chubby in her *Live!* performances for her role as a Latina in the mainstream to be recognized.

However, equally limiting are the assumptions made by the parents who seek to erase the differences of the body in the type of "color blindness" that is also often part of the mainstream discourses of diversity. Whereas parents who are concerned with Dora's brownness and chubbiness suggest that it is their children who will have the problem, similarly parents who advocate erasing the bodily differences also invoke the children's enjoyment as part of their argument. For example *TRDARC* states:

> Seriously and I mean this in the nicest way possible . . . you people are looking WAY to into this. You are not the ones that enjoy the show and are learning from it. You can't tell me that a toddler is going to notice the difference. All you are teaching your children to do is label people. Any child is going to remember the experience of seeing Dora in person and experiencing the atmosphere way before they remember if the live Dora had a true latino heritage. Let the children go and enjoy what life offers. (sic) (NickJr.com, 2004)

In one sense this poster is correct, in that children's enjoyment of and learning from the media do not necessarily have to come from easily recognizable ethnic or even human categories: witness Big Bird and Barney as prime examples of the latter. However, what is also true is the cumulative, long-term effect of watching the same types of peoples, the same stereotypes, in the collective psyche of the general population. For, lest we forget, one of the major reasons we worry so much about what the children are watching is that they will eventually become the adults who ostensibly will be responsible for the next generation.

In the same vein as *TRDARC* above, *Jbrma* on the "Dora Live: 'Who's the White Skinny Girl??'" thread states "Hey did your child like the show? Answer yes, then why does it matter" (NickJr.com). Answers such as these tend to emphasize a type of personal enjoyment of entertainment that is delinked from any awareness of the political, long-term, cumulative effects of representation. *hmicaela* appears on this "Dora Live Show-ethnicity?" thread twice, and her answers are particularly instructive in this regard. On the one hand *hmicaela* is one of the few people who acknowledge the ethnic breadth of Latinidad. She states that Latinas/os cover a range of skin colors, and she identifies her own children as fair Latinos. However, she also agrees with *TRDARC* stating that she does not care about the ethnicity of the actress who plays Dora:

> I don't care what the actress' ethnicity is, I was just answering the question. And you are right. I bet they picked an actress that could sing, dance and has childlike appeal. That is so much more important.
>
> BTW, I think I've seen a white girl as Jasmine at Disneyland. And I'm pretty sure their Pohcahontis is not Native American (though they get a litte more "tanned" person.) (sic) (NickJr.com, 2004).

Her comment places these discussions right in the middle of the economic and political structures that influence representation as well as the discourses of color blindness that seek to keep those structures invisible. Though *hmicaela* emphasizes both the audience's personal response to entertainment and the actress's personal responsibility to be qualified enough to get the job ("I bet they picked an actress that could sing, dance and has childlike appeal"), her invocation of the mega-corporation Disney must surely draw our attention to the intersections of capitalism and racism that shape and color mainstream media representations. So whereas *hmicaela* is correct that acting is about performativity and it should not matter what race or ethnicity the actor has as long as through representation s/he can perform that role, this also can be and is an issue of labor—if white actors can play any ethnicity, then the opportunity for actors of color diminishes. For it is not dark actors who are asked to play white people but quite the reverse! Second, these are not just personal issues that can be reduced to an individual response. These are institutional issues that we need to address as a culture.

Dora and the authentic language

One of *hmicaela's* other responses on the "Dora Live Show- ethnicity?" thread perhaps helps us understand further the ways in which Latina cultural authenticity is constructed within *Dora*. We have already outlined the types of contestations taking place over the "authentic" Latina geography and the "authentic" Latina body. However, some of the most apparent boundary building in *Dora* takes place around the issue of language usage. Perhaps *hmicaela* signals why this is the case when she states that the Latino community "is the only people group that is based on language rather than place of ancestry" (NickJr.com, 2004). Language, or rather Spanish, it would seem, is to be taken as the only valid signifier of Latina/o cultural identity. There are a number of threads on the Web board that deal with Spanish use. Most of the posts seek clarifications on the pronunciation, proper usage, and spelling of Spanish words. This is only to be expected given that there is quite a variety of Spanishes in the Americas. There is even a call for a Spanish Advisory Board from Web board participant *FJHDZ* who states that many of the Spanish words used in the program are "totally wrong and not pronounced properly" (NickJr.com, 2004). Though there is recognition that there are many different types of Spanishes, there remains the assumption that Spanish is the authentic marker of Latina/o identity.

One of the longest and most passionate threads on the board is the "why is Dora teaching Spanish?" thread, started by *asister*. *asister* says she is concerned that her little sister is learning Spanish on *Dora*, "We are in America so why don't they have programs to teach the Spanish kids English . . . We are in our very own country if people don't understand us, they should make the effort to, not us making the effort to understand them" (NickJr.com, 2004). Of the twenty-five responses that this post received only one agreed with *asister*. Nevertheless, it is in the respondents' support of the use of Spanish and the principles of diversity and multiculturalism that we actually see the insidious ways in which the hegemonic discourses of whiteness continue to assert themselves. There were a number of different reasons advanced by the Web participants seeking to challenge *asister's* perspective, however, by far the majority of responses were couched in the language of education and employment possibilities. Thus a number of the people on the Web board suggested that learning a second language would expand the children's brain development. Additionally many board members drew reference to the changing demographics in the United States that make it imperative for children to learn Spanish if they hope to get a job in the future. This type of reasoning was also present in the Nickelodeon's justification for teaching Spanish, and also present in these justifications is the continuation of the idea that Latinas/os are foreign as marked by their use of a foreign language. For example co-creator Valerie Walsh in explaining why Spanish is used in the show lays most of the emphasis on Spanish being a *secondary* language:

> Educators believe that introducing a second language to a child before the age of 6 or 7 is
> an important factor in his/her ability to achieve fluency. For many of our preschool viewers, Dora is their first encounter with *a* foreign language. As such, the show might teach them
> a little Spanish and make them curious and interested in learning more, or simply make them

aware of and comfortable with foreign languages. For our Spanish-speaking preschool viewers, seeing Dora use Spanish might encourage them to take pride in being bilingual. (Walsh, 2004)

Even Walsh's acknowledgment of the Spanish-speaking viewers appears to problematize Spanish. In other words, Spanish cannot be afforded the taken-for-granted status of English, it can only be viewed within a framework that accepts it as a resistant Other. In fact, the last sentence invokes the classic language of acculturation in terms of ethnic populations in the media in which issues of language and pride have to be bracketed into ethnic identity while privileging and centering adoption of mainstream culture.

Web board posts that spoke to these issues tended to wittingly or unwittingly suggest that the changing face of the United States was a threat to "Americans'"—taken to mean Anglo-Americans'—rightful positions of control. For example *wdr-wmn13* states, "Obviously you don't get out much. I live in Houston, TX and I can't find a job because they want Bilingual speaking people this does outrage me as a American English speaking person . . ." (NickJr.com, 2004). *flemaer* is even more protective of his/her privilege and resentful of perceived incursions into that privilege by speakers of Spanish:

I think Spanish is a very important tool. Here is why. I could have had my dream job about a year ago. I lost it. I was more qualified than the person who got it. They went with the other person because she could speak Spanish, and a lot of the clients speak Spanish. I have spent the last year in school learning Spanish. I am now an assistant to the person who has my dream job. If only I had learned Spanish before. I would have the job I want, and not just be an assistant. (NickJr.com, 2004)

Note that both these writers are challenging *asister*'s contention that "American" children should learn English; however, they are still using the same frame of reference that positions English speakers as "Americans" and Spanish speakers as a foreign threat. So effective is this binary that the writers are unable to recognize when their logic breaks down. Thus *flemaer* continues to contend that he/she is more qualified than the person who got the "dream job" despite the fact that speaking Spanish was a key qualification and that *flemaer* lacked that qualification. Learning Spanish, therefore, in instances such as these becomes a tool to ensure white privilege in the face of the Latina/o presence that is understood as being a foreign presence.

Discussion

Dora is the embodiment of synchretic tropicalizing discourses. Her brownness, situated in an unspecific tropical space, is emblematic of the commodification of this new constructed audience that we now know as Latina/os. The fact that her pet,

Boots, is a monkey draws on a tradition of Latinas in the U.S. mainstream media. Frida Kahlo's "Self-Portrait with Monkey" (1938), a widely circulated image in the U.S., sets the stage for a little Latina girl having a monkey as a pet. Another contemporarily prominent Latina girl in the mainstream, Josefina of the American Girl series, also has an unusual pet, a goat. Of course, this becomes yet another element of difference vis-à-vis whiteness. Moreover, both her English-dominant use of language, with a little bit of Spanish sprinkled throughout the series, as well as her independence and command of computer technology situate her within the synchretic discourse of Latinidad. She embodies the modern brown Latina.

Simultaneously Dora, as a cybercitizen of the technoscientific U.S., inherits the subjectivity of tourist and consumer of the world through the window of the computer. As Nakamura (2002) writes: "The enraptured American (sic) schoolchildren, with their backpacks and their French braids, are framed as user-travelers" (p. 92). Although she is writing about a Compaq advertisement, the statement is eerily applicable to *Dora the Explorer*. Although a girl, and a brown one at that, Dora still embodies the all-seeing, exploring, touring Western subject that is so prevalent in contemporary discourses of the Internet (Haraway, 1997). That is, in relation to the U.S. imaginary of itself as White, she is exotic, yet in relation to the rest of the world, she is the rational, technologically enabled, normative subject. Her embodied presence in the Internet retains both tropicalist and Eurocentric imperialist traces. Whereas Dora signs in as a third world citizen in a first world media diet, her relationship to the global is certainly that of a first world subjectivity relying on "stable, iconic images of the other" (Nakamura, 2002, p. 92).

Yet this book is about girls on the Internet, and our investigation centered on the discussions about Dora's subjectivity on Internet chat streams about the television show and the touring live show performances. In particular, we challenge the construction of the Internet as a location of disembodied subjectivity, at least for ethnically marked peoples. Whereas white subjects may indeed participate in Web chats about themselves in a disembodied manner, which allows for all sorts of possibilities including gender and ethnic cyber cross-dressings, the *Dora* chat rooms included, and sometimes demanded, an embodied and re-territorialized subjectivity precisely to police the boundaries between Latinidad and whiteness. Both from within Latinidad, or at least in terms of how posters identified themselves in the chat rooms, and without, many participants demanded a clear tale of geographic origin, an identifiable Latina body, and Spanish-language authenticity. On the issue of language, despite disagreement as to the benefit of a second language and its place and value in U.S. culture, posters write from a standpoint of undeniable English-language centrality and Spanish as a minor component of a competitive, workplace edge and/or of an acculturated ethnic identity. Additionally, explicit traces of resentment surface in these chat rooms, wherein English, non-Spanish speakers assert their centrality in U.S. culture through their attempt to marginalize bi- or multi-lingual people.

In fact, as the official Nick Jr. Web site tells us, the importance of *Dora*, the television show, the live performances, and the associated synergistic products is not to include another ethnic media entry show into the U.S. mainstream. Nor could one argue that *Dora's* Internet presence is an instance of disembodied subjectivity. Rather this character, in her myriad manifestations, is a graphic reiteration of tropicalist tropes. *Dora's* tropicalization serves to commodify Latinidad and maintain a white Euro-centric "we" or target audience through the inclusion of others who must not forget they are others—not even in the so-called disembodied space of the Internet.

Notes

1.   We must intervene here against the abuse of the word "interactive." The fact is that the show is reactive, as children's responses remain within their viewing environment and do not return to the show's producers.
2.   For the rest of this paper, participants in the chat groups will be referred to by the names they use on the Web board.

References

Aparicio, F. R., &. Chávez-Silverman, S. (Eds.) (1997). *Tropicalizations: Transcultural representations of Latinidad*. Hanover, NH: University Press of New England.

Appadurai, A. (1996). Diversity and disciplinarity as cultural artifacts. In C. Nelson, & D. P. Gaonkar (Eds.), *Disciplinarity and dissent in cultural studies* (pp. 23–36). New York: Routledge.

Barney, C. (2002, August 3). Kids' channels beat networks to diversity. *San Diego Union Tribune*. Retrieved February 18, 2004, from Lexis-Nexis.

Bauder, D. (2003, May 22). TV diversity showing improvement. *Contra Costa Times*. Retrieved February 18, 2004, from Children Now.

Butler, J. (1990). *Gender trouble: Feminism and the subversion of identity*. New York: Routledge.

Dávila, A. (2001). *Latinos Inc.: The marketing and making of a people*. Berkeley: University of California Press.

Denzin, N. (1999). Cybertalk and the method of instances. In S. Jones (Ed.), *Doing Internet research: Critical issues and methods for examining the Net* (pp. 107–123). Thousand Oaks, CA: Sage.

Felski, R. (1995). *The gender of modernity*. Cambridge, MA: Harvard University Press.

Fusco, C. (1995). *English is broken here: Notes on cultural fusion in the America*s. New York: New Press.

Gifford, C. (2004) Meet the creators. Retrieved February 2, 2004, from http://www.nickjr.com/home/shows/dora/meet_doras_creators.jhtml

Giroux, H. A. (1991). *Border crossings: Cultural workers and the politics of education*. New York: Routledge.

Hacker, K. L., & Steiner, R. (2002). The digital divide for Hispanic Americans. *The Howard Journal of Communications*, *13*, 267–283.

Halter, M. (2000). *Shopping for identity: The marketing of ethnicity*. New York: Schocken Books.

Haraway, D. J. (1991). *Simians, cyborgs, and women: The reinvention of nature.* New York: Routledge.

Haraway, D. J. (1997). *Modest-Witness@Second-Millennium.FemaleMan-Meets-OncoMouse: Feminism and technoscience.* New York: Routledge.

Kolko, B. E, Nakamura, L., & Rodman, G. B. (Eds.). (2000). *Race in cyberspace.* New York: Routledge.

Latina/o communication studies. (2004). *Communication Review, 7*(2).

Lemish, D., Liebes, T., & Seidmann, V. (2001). Gendered media meanings and uses. In S. Livingston, & M. Bovill (Eds.), *Children and their changing media environment: A European comparative study* (pp. 263–282). Mahwah, NJ: Lawrence Erlbaum.

Levine, E. (2001). Constructing a market, constructing an ethnicity: U.S. Spanish-language media and the formation of a syncretic Latino/a identity. *Studies in Latin American Popular Culture, 20,* 33–50.

Livingston, S., & Bovill, M. (Eds.). (2001). *Children and their changing media environment: A European comparative study.* Mahwah, NJ: Lawrence Erlbaum.

Mahalingam, R., & McCarthy, C. (Eds.). (2000). *Multicultural curriculum: New directions for social theory, practice and policy.* New York: Routledge.

Martinez, J. (2002, January 8). Impact of Latinos in 2001. *The Denver Post.* Retrieved February 18, 2004, from Lexis-Nexis.

Marvin, C. (1988). *When old technologies were new: Thinking about electric communication in the late nineteenth century.* New York: Oxford University Press.

Mastronardi, M. (in press). Policing dis-order: Moral panic and pro-Ana citizenship. *Cultural studies: Critical methodologies.*

Mayer, V. (2003). *Producing dreams, consuming youth: Mexican Americans and mass media.* Brunswick, NJ: Rutgers University Press.

Mayer, V. (2004). Please pass the pan: Re-theorizing the map of panlatinidad in communication research. *Communication Review, 7*(2), 113–124.

McCarthy, C. (1998). *The uses of culture: education and the limits of ethnic affiliation.* New York: Routledge.

McLaine, S. (2003). Ethnic online communities: Between profit and purpose. In M. McCaughey & M. D. Ayers. (Eds.), *Cyberactivism: Online activism in theory and practice* (pp. 232–252). New York: Routledge.

Mohanty, C. (2003). *Feminism without borders: Decolonizing theory, practicing solidarity.* Durham, NC: Duke University Press.

Nakamura, L. (2002). *Cybertypes: Race, ethnicity, and identity on the Internet.* New York: Routledge.

Navvaro. M. (2001, July 2). Nickelodeon's bilingual cartoon *Dora* is a hit. *New York Times.* Retrieved February 18, 2004, from Lexis-Nexis.

NickJr.com http://www.nickjr.com/home/shows/dora/index.jhtml

Rickert, A., & Sacharow, A. (2000). *It's a woman's World Wide Web.* Media Matrix, Jupiter Communications. Retrieved March 2, 2004, from http://pj.dowling.home.att.net/4300/women.pdf

Sangha, J. K. (2004, February). How race is reproduced in cyberspace: Understanding the privileging of Whiteness in online forums. Paper presented at the University of Ottawa Law School Conference on Violence and the Internet, Ottawa, Canada.

Seiter, E. (2002). New technologies. In T. Miller (Ed.), *Television studies* (pp. 34–37). London: BFI Publishing.

Stern, S. R. (2002). Virtually speaking: Girls' self-disclosure on the WWW. *Women's studies in communication, 25*(2), 223–248.

Suden, J. (2001). What happened to difference in cyberspace? The (re)turn of the she-cyborg. *Feminist Media Studies, 1*(2), 211–232.

This week in Chicago. (2004, February 13–21). *167*(10): 6.

U. S. Department of Commerce (2002). *A nation online: How Americans are expanding their use of the Internet.* Retrieved January 28, 2004, from http://www.ntia.doc.gov/ntiahome/dn/

Walsh, V. Meet the creators. Retrieved February 2, 2004, from http://www.nickjr.com/home/shows/dora/meet_doras_creators.jhtml

6

You're Sixteen, You're Dutiful, You're Online

"Fangirls" and the Negotiation of Age and/or Gender Subjectivities in TV Newsgroups

Christine Scodari

Introduction

As the millennium dawned, "the explosive growth in Internet usage among teenage girls" allowed females to overtake males in their utilization of computer-mediated communication (Regan, 2000). Until then, women were positioned on the shy side of a digital divide that persists in terms of age, race, class, and other variables. Particular online sites, however, can attract users with traits that do not mirror Internet users as a whole. Participation in cyber-communities that coalesce around cultural texts original to another medium—in this case, television—is skewed both in terms of who is in the initial audience and which members of that audience are likely to have the access, ability, and inclination to use the Internet to indulge their interest. As a result, teen girls' association with these sites may exceed, match, or fall short of their overall presence online. Sites devoted to texts that appeal to a cross section of viewers create a context in which the negotiation of identity—a momentous exercise for adolescent females—is particularly revealing. This chapter addresses the following issue: How do teen girl fans of television shows and genres with diverse audiences assert and negotiate subjectivity and identity in the context of computer-mediated deliberation of these cultural products in Usenet newsgroups?

Interdisciplinary Groundwork

Over the last three decades, germinal works addressing the psychology and self-esteem of adolescent girls have exposed the "silencing" that can occur during this pivotal period of development (Chodorow, 1978; Gilligan, 1982; Pipher, 1994). Females have traditionally been taught to value themselves primarily in the context of interpersonal relationships, and, consequently, overt expressions of unique identity that might ruffle such relationships decline as girls navigate beyond puberty. In addition to provoking much political debate, these works have prompted further academic studies (Brumberg, 1997; Way, 1998) as well as popular offerings (Haag, 2000; Nichter, 2000) that allow girls' voices free rein. But is giving voice to girls enough? Brumberg (2003), considering the less scholarly treatments, observed:

> Although adolescent "voice books" are a deserving corrective to the ways in which adolescent girls have been silenced and overlooked, they are obviously part and parcel of an entrepreneurial and confessional culture that can be lurid and tawdry. That culture, more than contemporary feminism, explains the emotional tone and the highly individualistic politics of such books. (p. B7)

Accordingly, this study is particularly sensitive to whether the identity negotiations it unearths venture beyond such individualistic politics.

In the provinces of cultural and media studies, girls' identities and cultural collaborations are the foci of the essays assembled in Mazzarella and Pecora's (1999) *Growing Up Girls* as well as *Feminism and Youth Culture*, a compilation of McRobbie's (2000) canonical works from the 1970s and beyond. McRobbie's classic analysis of *Jackie*, a British magazine for teen girls, discerned a subjectivity of "romantic individualism" that is relevant to the research undertaken here and by which hegemonic cultural identities of this group become amplified (p. 69). In manifesting this subjectivity, *Jackie* asserted

> a certain class-less, race-less sameness, a kind of false unity which assumes a common experience of womanhood or girlhood. By isolating a particular phase or age as the focus of interest, one which coincides roughly with that of its readers, the magazine is ascribing to age, certain ideological meanings. Adolescence comes to be synonymous with *Jackie's* definition of it. (p. 69)

Jackie animated four codes of romantic individualism, also essential to the present study: (1) romance; (2) beauty and fashion; (3) personal and emotional life; and (4) pop stars and music. These ideas are consonant with McRobbie's (2000) ethnography of British working class girls and the psychological conclusions of Chodorow (1978), Gilligan (1982), and others. While the McRobbie (2000) and Mazzarella and Pecora (1999) volumes remedy long-standing neglect of girls as social actors, and probe their negotiation of identity in relation to more traditional cultural channels and forms, they do not confront the comparatively new medium of the Internet.

Book-length investigations of women's Internet usage include Shade's (2002) *Gender & Community in the Social Construction of the Internet* and Consalvo and Paasonen's (2002) *Women & Everyday Uses of the Internet*, both of which consider broader issues of identity but pay scant attention to girls and/or age difference. Only Oksman's (2002) essay in the latter volume addresses Finnish girls' participation in "Virtual Stables," a fantasy game about caring for horses, as a counterpoint to more violent, less pedagogical online games for boys. Elsewhere, Stern (2002) examined the home pages of young girls and unearthed highly personal narratives seeking to establish a sense of identity and place in primarily private sphere contexts.

My own research exploring dimensions of age and/or gender in the online responses and negotiations of television fans serves as grist for this inquiry. Departing from previous studies of gender and online fandom such as Baym's (2000) analysis of interaction among soap opera fans on Usenet, I interrogated the claim that Internet communities minimize, especially among women, differences that exist offline. Teen girls were among the fans of *The X-Files* who used the Internet to lobby unreceptive creators and against their favored audience—young males—on behalf of romantic continuity in the depiction of the relationship between the lead characters, Agents Mulder and Scully (Scodari & Felder, 2000). Soap opera "fangirls," on the other hand, are strategically pursued and catered to by producers. Unaware or unconvinced of their exalted position, however, these cyber-fan subjects were flummoxed and placed on the defensive by older participants' claims of marginalization and devaluation (Scodari, 2004). The interpellative pull of younger female characters in both soaps (Scodari, 2004) and science fiction series (Scodari, 2003) can put teen girls at odds with the corresponding subjectivities and penchants of their older and/or male counterparts as Web deliberations unfold.

History, terms, and concepts pertaining to computer-mediated communication, especially in regard to Usenet newsgroups, also demand attention. In the early days of the Internet, Usenet became a primary mechanism by which asynchronous interaction among diverse, geographically dispersed users occurred. Such access could be facilitated via "news reader" applications. Bulletin board, forum, and/or message board opportunities were also available, mostly through suppliers such as Prodigy, CompuServe, and AOL. Increasingly, throughout the early to mid-1990s, Web communities evolved to accommodate asynchronous (bulletin board) and, later, synchronous (chat) interactions, and providers featured portals to the Web, often in addition to subscriber-only services. Discursive options multiplied, resulting in a millennial decline in Usenet participation that occurred after many academic inquiries specific to Usenet, such as those by MacKinnon (1995) and Baym (2000), took place. This decline has, perhaps, varied the degree and diversity of involvement in the groups to be studied here.

Net presence, defined by Agre (1994, p. 1) as "the continuing fact of being metonymically 'present' through various net activities," is also pertinent to subjectivity and assertions of identity. In addition, Barnes (2003) catalogues tools of online identity construction such as personal profiles created by participants and recorded on Web sites, signature files that automatically attach contact information and other personal tidbits to each online communiqué, and screen names adopted by users. In relatively heterogenous groups in which common perspectives cannot be taken for granted, making one's place is seen as especially challenging and telling, and participatory fluxes and customary identity apparatuses are key factors.

In order to assess the ways and means by which teen girls seek and negotiate subjectivity, this study's data were winnowed from records of online interaction that occurred in the relatively recent past and then subjected to non-participatory observation and analysis. Usenet newsgroups for TV series with which this researcher is familiar, and which were likely to attract a significant number of teen enthusiasts, were pinpointed for scrutiny. These included those devoted to CBS's *Survivor* (alt.tv.survivor), FOX's *The X-Files* (alt.tv.x-files), NBC's *ER* (alt-tv-er), ABC soaps (rec.arts.tv.soaps.abc), CBS soaps (rec.arts.tv.soaps.cbs), and NBC and all other soaps (rec.arts.tv.soaps.misc). A search was conducted via the Usenet archive (groups.google.com) for messages in which the sender declared, "I'm 15," "I'm 16," "I'm 17," or their equivalents. Posters who might not be eighteen by the publication date of this essay were exempted. A search for all posts to the relevant newsgroup by each of these participants was then performed. Results were initially perused in order to ascertain gender in the use of signatures, pronouns, and other indicators. The first posters presenting as female, up to a maximum of six per group, formed the subject pool. Then, returning to the results for each subject, and moving backward and/or forward in time up to four screens from the one containing the age declaration, substantive messages, up to a maximum of seven per subject, were culled. Those dated when the subject could have been younger than thirteen or older than eighteen were discarded. Overall trends and tendencies in the data were identified and assessed. Finally, messages were categorized using Glaser and Strauss's (1967) constant comparative method and interpreted via pertinent theory.

Negotiating Subjectivities

While McRobbie (2000) investigated *Jackie*'s efforts to carve out and define its teen girl consumers through a subjectivity of romantic individualism, the "fangirls" observed for this inquiry responded to TV texts that are not geared exclusively or, in some cases, primarily, to them. Consequently, the "false unity which assumes a common experience of womanhood or girlhood" (p. 69) that is circulated by *Jackie* and other dominant teen girl culture artifacts might be suppressed or invigorated in light of the broader participation evident. Uncovering associations or dissociations with television texts through a second level of identification—that is, in terms of their creators, stories, performers, and/or characters—can offer more penetrating analysis in this context, perhaps exposing the extent to which romantic individualism is embraced or challenged by subjects. As Gledhill (1988) has noted, negotiation can occur at the levels of production, text, and consumption, and fans' negotiations of texts are often influenced by the subjectivities that these texts, via specific plots, stars, and personae, prefer or allow.

Multi-level identification is conspicuous among the following categories of postings, which were ascertained using the constant comparative method: (1) character/plot identifications and/or disidentifications; (2) attitudes toward producer/creator decisions; (3) defenses of teens; (4) statements of devotion to shows, characters, and/or actors; and (5) tangential comments about related interests, everyday life, etc. Some posts addressed more than one of these themes and, in many instances, hybridization with one or more of McRobbie's (2000) codes of romantic individualism was apparent. In addition, messages often inserted and responded to remarks made by other users, thereby constituting negotiation between and among fans. Wherever appropriate, these other users are sparingly quoted and/or paraphrased in light of subjects' observations and negotiations with the text.

The search of the *Survivor* newsgroup generated no subjects. One subject was identified from the miscellaneous soap opera newsgroup, two from the CBS soap opera newsgroup, five each from *X-Files* and *ER* newsgroups, and six from the ABC soap opera newsgroup. The largest number of messages in the final data set emanated from the ABC soap opera and *The X-Files* newsgroups. The earliest messages were dated 1995 and the latest were posted in 2002. Eight subjects produced no substantive posts beyond the age declaration. In the course of analysis, subjects are not identified with actual or screen names. Those with more than a single post in the data set, however, are referred to by pseudonym.

The categories spurring the age declarations are noteworthy. Several declarations were made in response to newsgroup "polls" or solicitations of user age and/or other demographics. Of the declarations that were otherwise substantive, most highlighted character/plot identifications or disidentifications. Next were messages that mainly defended teens or teen girlhood in rebuttal to negative insinuations or stereotypes. All other categories lagged behind in this regard.

Signatures and signature files, some of which reveal allegiance to fan Web sites and/or sub-groups, often reflect Net presence and assert identity within the fan community. For example, subject Mandy, an *X-Files* devotee, advertised that she was "X-Ville's official Auto Mechanic and hopeful waitress." Shannon, an avid supporter of *X-Files* star David Duchovny, declared herself to be a member of SPCDD (Society for the Prevention of Cruelty to David Duchovny) and quoted from a fan fiction passage describing his character, Agent Mulder, as "damaged goods." As fans of ABC's *General Hospital*, Sandi listed "FGC [Favorite Guy Character] Lucky, Keeper of Broken Rules and Mended Fences" among other attachments and roles in her signature lines, while Noelle announced: "MHGC [Most Hated Girl Character] Lily's voice." *ER* enthusiast Lynne professed to be a "science fiction junkie." As we shall see, such proclamations often predict the type of issue that is most likely to provoke participation on the part of these subjects as well as the intensity of their comments.

Character/Plot Identifications and/or Disidentifications

Many of the character/plot identifications or disidentifications involved romantic storylines, in accordance with romantic individualism and the code of romance unpacked by McRobbie (2000). One subject, a fan of *The Young and the Restless* (CBS), responded to another user's query about which of "two hot hunks o' boy flesh" a young heroine, Mac, should favor:

> I'm 17 years old right now, and I can tell you that I wouldn't even think twice about choosing Raul. A teenage girl goes through a lot of pressure at this age, and I can't imagine having a boyfriend like Billy would really help. Raul is such a good influence on Mac, after all she's been through. He's showing her love. Billy could show her how to chug beer.

As McRobbie (2000) observed of the stories found in *Jackie*, male heroes, while more independent than heroines, defied sociological expectations by being keen on romance rather than sex. This user suggests that factors other than the mere attractiveness or "coolness" of a male character might determine his appeal through the eyes of a heroine with whom a teen girl identifies. Her negotiation, primarily with the text and via this female character, references teen girlhood as a unique identity entitled to independent agency, as consistent with romantic individualism. Her "independent" choice, however, is the more respectable one—a preference which appears to be privileged by the text.

Shannon, the aforementioned fan of *The X-Files*, David Duchovny, and his character, pondered another user's speculation that Agent Mulder might turn to a prostitute "for release":

> Though Mulder somehow seems far too shy to ever do such a thing, if he did, I would pity him so badly, and could perfectly easily see him going to her if only for company, if not for sex. I mean, with all intentions of getting his $300 dollars worth (I'm sixteen and female,

forgive me if my price range is uneducated), but instead ending up in a puddle of his own misery.

Unlike the *Young and the Restless* fan quoted above, Shannon admits her age and gender in order to plead less rather than more familiarity with a certain aspect of this subject matter. Her role as protector of not only the reputation of the actor but of his character's interpersonal esteem, is noteworthy, and once again harmonious with McRobbie's (2000) romantic codes, insofar as many of the male leads in *Jackie*'s short stories aroused "maternal feelings in girls" (p. 84). This fan adopts a subjectivity not already taken up by a character, instead negotiating a position within the text as Mulder's hypothetical caretaker.

Tori, an *X-Files* fan generally more sympathetic to Mulder's female counterpart Agent Scully (as played by Gillian Anderson), reacted at length to an episode entitled "First Person Shooter." While some in the group argued that it pandered to men and demeaned women, Tori was philosophical, explaining that it made her feel "sorry" for males who viewed their gender as depicted in the episode. Reacting to the episode's "bombshell" character, she disclosed:

> While others of my age and gender (I'm 16 and proud to be home to two X chromosomes) might have been intimidated by Maitraya/Afterglow, I don't think I am. I've reconciled myself to the fact that I'll never be one of THEM (thanks to circumstances beyond my control that have caused some scars on my face and body). . . . Beauty doesn't last forever. Neither do scars.

The circumstances Tori references clearly accelerated and heightened an inevitable struggle to come to terms with body image, a major identity issue for most teen girls, and a key facet of McRobbie's (2000) code of fashion and beauty as constituted in *Jackie*. Whereas *Jackie* encouraged conformity to beauty norms, Tori appears to renounce them. However, having negotiated her own individualistic solution, she rebuffs the collective outcry of other fans.

In a post that also defended teens as "smarter than all adults would like to admit," Renée, a *General Hospital* fan, elucidated that they "understand what adults don't seem to get, that sex is not love and love is not sex." Consequently, she perceived teen heroine Robin as too "immature and stupid" to be in a serious relationship with her boyfriend, proclaiming: "I feel like barfing when she starts talking about her love for Stone." Renée's dissociative textual negotiation embraces both traditional values regarding sex and a unique teen girl identity.

Gina, a fan of ABC's *All My Children* among other soaps, admitted to liking older characters more than younger ones. However, of middle-aged Brooke, she remarked: "I'm really enjoying Brooke these days. I never would have said that during the Pierce days or the horrible mourning period that followed. I like this new Brooke who is slightly wacko and who really wants to get laid." *Liking* in this instance appears to refer to savoring Brooke as a caricature—a desperate harpy—

rather than sharing her romantic subjectivity. This mirrors tendencies unearthed by this researcher's earlier exploration of gender and age as factors in cyberfan nego-tiations of soap opera romance (Scodari, 2004). Younger fans tended to appreciate older characters by positioning themselves outside the text rather than inserting themselves within it.

Debates over favorite characters, couples, plots, and types of plots also distin-guished this category. When another fan complained that few in the *ER* group stood up for the romantic coupling of Doug and Carol as they faced a crisis in their rela-tionship, Lynne, acknowledging the petty factionalism that often pervades fan communities, while failing to transcend it, protested: "Hey, I consider myself a Doug/Carol groupie, but you're all alone on this one. . . . You complain about too many couples, but are unwilling to give up on your couple." Here, the negotiation is between fans, and the implication is that when it comes to romance, it's every woman for herself.

General Hospital fan Noelle expounded on exaggeration in soap opera romance:

> Don't soaps want to have hot romances that heat up the screen? A couple fighting to be together? I mean, sweet, "friendshipy" marriages might be nice in real life, but TV romances need something more. Luke and Laura are an example of "the big love," and look how pop-ular they became.

Interestingly, Luke and Laura were a couple whose genesis in an instance of rape has been hotly debated (see Hayward, 1997; Nochimson; 1992; Scodari, 2004). Luke is perhaps the most extreme example of the "bad boy" romantic hero who, accord-ing to McRobbie (2000), "the girl must tame" (p. 84). Evident in earlier soap opera study (Scodari, 2004) and analyses of romance novels (Barlow & Krentz, 1995; Radway, 1984), this archetype again supports the notion that each woman is on her own in establishing and maintaining an identity forged in and by romance.

As previously mentioned, a perennial debate in the annals of *X-Files* fandom was whether Agents Mulder and Scully should evolve into a romantic couple. This controversy, weighed in prior research (Scodari & Felder, 2000), is also apparent here. Unlike many others of her gender and age group, Pam insisted that although the agents should hook up eventually, if they "get together before the last season and finale episodes of the show . . . then the show will be ruined FOREVER!" Similarly, Mandy argued that "M&S [Mulder and Scully] should only jump each other in the final final episode," but also that "a make-out scene, whether intentional or not, would be really cool." However, even the prospect of a post-series romance was adamantly opposed by the preferred audience of somewhat older, mostly male fans who, unlike the heroes of the magazine stories McRobbie (2000) examined, tend-ed to be a decidedly unromantic lot (Scodari & Felder, 2000). A subjectivity of romantic individualism, therefore, is a domain of women, to be negotiated in terms of the text and its characters even if forsworn by actual men.

Attitudes Toward Producer/Creator Decisions

Much teen fan commentary on producer/creator decisions and directions considered issues important to teen girls such as body image as well as the wisdom of catering to teens. An *ER* fan inveighed against a code of fashion and beauty (McRobbie, 2000) in responding to a controversial casting choice: "I'm fine the way I am. I'm a size 14. I already have my mom and clothing companies telling me I'm fat. I don't need a casting director telling me I'm fat. Hello, *ER* producers. I can play Rachel!" Gina, a fan of several soaps, said this about "teen storylines":

> It seems to me that many of these storylines are entirely too forced because TPTB [the powers that be] feel the need to attract younger viewers. I recently went away for the weekend with hundreds of high school seniors . . . and we all moaned about missing our Friday soaps. Of all the storylines we discussed and got excited about, not a single one involved the teens.

Earlier investigation concluded that teens may not always gravitate toward characters their own age (Scodari, 2004). Their identification is mostly aspirational; that is, they are more likely to follow the stories of characters in their twenties. As a previous quote from Gina also attests, middle-aged and older characters tend to be intriguing to younger viewers as objects of humor—an appeal intentionally cultivated by *Days of Our Lives* (NBC), among other soaps.

Yet some teen girls, such as this fan of *Days of Our Lives*, reacted in unexpected ways:

> The damn thing is written by fourth graders. The last straw was realizing that the little girl Lucy was the only believable character I have seen. . . . She was a better actress than all the adults put together. What I wouldn't give for a chance . . . to take one day's script and revise it. If the writers have any guts, I dare them to write me. I'm 17 and literary editor of my school's literary magazine.

Katie concurred, parodying the acronym for *Days of Our Lives*: "I've watched DrOOL . . . since I was born because my whole family is addicted, but because that show has completely disgusted me with slow, repetitive storylines, I'm switching to *All My Children*." Discontentment, therefore, can be managed by switching to another product, or even by imagining oneself, the individual fan, as a "savior" usurping creators' roles.

When controversy arose with respect to the casting of racial minorities in *The X-Files*, Tori acknowledged:

I think casting coordinators are less willing to hire minorities for lead roles because, for most shows and films to succeed, there has to be someone for the audience to identify with, and I think a lot of white people just have trouble empathizing with a minority character.

This is an astute comment that, as might be expected, does not implicate deeper structures of mass media operation. In similar terms, fans of this series had been aware that the FOX network initially disapproved of the casting of Gillian Anderson as Agent Scully, preferring an overtly and stereotypically "sexy" type for the role (Scodari & Felder, 2000). This helped to kindle sporadic rumors that she might be dropped against the wishes of the show's creator, Chris Carter. When a new female character began appearing on a recurring basis in the fourth season, just as Carter began paying more attention to a fresher program in his stable, Mandy lamented: "Oh my gosh, I'm praying that isn't true. If GA [Gillian Anderson] leaves the show it is doomed. It's that damn *Millennium* that is ruining everything." In the same vein, Hannah defended a lighthearted episode of *The X-Files* but bemoaned what she and other fans saw as a trend away from the program's roots: "I really wish CC [Carter] would come up with some scarier episodes. I mean when the show first started he said his object was to scare the daylights out of millions of watchers. . . ." These critiques reveal how identifications with the text, its characters, and its performers gel in the context of defending and/or renouncing producers and their preferences.

Defenses of Teens

Forceful expressions of identity, many tapping into the code of personal and emotional life probed by McRobbie (2000), frequently surfaced in concert with defenses of teens. *X-Files* fan Pam objected to another poster's insinuation of a negative stereotype: "I guess I can say that I sort of resent that. I'm 16 and I do not have Junior High humor." Zoe, a fan of *Young and the Restless*, reacted as follows to an elderly poster's suggestion that teens should "have a life" and not "waste time" with soaps:

You tell me that I'm a bad person, when I volunteer my time, babysit, taking all HONORS AND AP (advanced placement, a.k.a. college credit courses) classes in my junior year, on high honor roll, #50 out of almost 500 people, and also in National Honor Society. . . . You're right I don't have a life cuz it is consumed by hours of homework. . . . Usually when I'm doing homework is when I have Y&R [*Young and the Restless*] and B&B [*Bold and the Beautiful*] on in the background.

Similarly, Hannah bristled when someone insisted that teens had no business watching *The X-Files*: "I'm really tired of everyone underestimating me, or trying to exclude me because I'm young. I'm 16 years old, and I think it's time that older generations got used to us, and learned that we have brains too." *ER* fan Lynne responded to other posters' suppositions about why a very young but much disdained character had been introduced:

> I really resent the comments that the writers are trying to appeal to teenagers and that's what's causing the downfall of *ER*. I'm 17 and found the episode just as boring as the next person. . . . The success of a show is dependent upon the type of people it appeals to, not their ages (yeah, yeah, I know the whole demographics thing, but the most coveted one is 18–49, not teenagers).

In a prior study of soap opera fans (Scodari, 2004), teens were observed to be similarly defensive when older viewers complained about shows indulging young audiences at the expense of longtime fans. Lynne not only rejects the premise that older viewers are marginalized, but the presumed result of such a strategy in the show's text, reflecting an apparent need to maintain alignment with older newsgroup participants, at least in terms of their critical judgments about the show's content. Yet, these censures of and rebuttals by teens attest to the culture's ascription of "certain ideological meanings" to age (McRobbie, 2000, p. 69).

Another young *ER* fan was infuriated when, after an episode about an ill and disabled teen, an older user suggested that teens were too "naive" to understand that they should not have a say in what medical procedures are performed on them:

> I'm extremely pissed off at this remark. . . . According to you, a child has no say in what happens to their body or their life. Children are too immature to even "move their arms or their legs," so *why* should they be able to be "let in" on what other people are deciding about their health?

Despite their defiant tone, and this last subject's repudiation of adult authority, the comments in this category advocate the "proper" traits and moral rectitude purveyed in *Jackie* (McRobbie, 2000) while displaying age-based identification.

Statements of Devotion to Shows, Characters, and/or Actors

Outright statements of devotion to shows, characters, and/or actors constituted another prevalent theme. Pam defended *X-Files* star Gillian Anderson against criticism related to publicity about her alleged "wildness" as a teen:

> Gillian, you GO girl! Ok, I don't understand why her past is such a biggie either. These same people who are making a big deal out of it should look at themselves as well. . . . All of you stop messing with her. The people that tell shows like Hardcopy about this crap are just jealous. Go get your own husband, baby, and tv show and leave her alone.

With the news in May of 2000 that Anderson's co-star, David Duchovny, was re-negotiating his contract and might not be returning to the show, Tori chimed in to laud Anderson's contribution to the program: "IMHO [in my honest/humble opinion] Ms. Anderson should be the one getting those hugeass offers and accommodations that Mr. Duchovny's getting." Shannon, on the other hand, had been more interested in supporting Duchovny: "Why do people bash on DD [Duchovny]? It only seems to be the people who seem to be all for GA [Anderson] and are always saying how horrible he is and everything. He's a human being! Get over it!"

The factionalism within this fan community is well documented (Scodari & Felder, 2000). McRobbie (2000) observed that the image of male pop stars profiled in *Jackie* depended on their being represented as unattached, in apparent contradiction to the modes of address activated by the magazine's fiction, designed to invoke reader identification with romantic heroines. In addition, the romantic codes manifested in *Jackie's* stories required that the heroine "fight to *get* and *keep* her man," according to which she could "*never* trust another woman" (2000, p. 85). In these respects, romantic individualism resonates in the attachments the fans in this study form with TV characters and, by extension, the actors who play them. In fact, previous inquiry (Scodari, 2003) attributed the "slash" fan fiction phenomenon, in which mostly female authors concoct homoerotic stories about male characters poached from one or more popular texts, to the desire of some women to romanticize their idols without having to contend or compete with other women, even if they are fictional. Young authors of "Mary Sue" fan fiction insert themselves into each of their stories as the central female figure. In seeking to invigorate the romantic code through a "slash" and/or "Mary Sue" mode of subjectivity, such a fan renders the object(s) of her devotion, like *Jackie's* pop star personae, as exclusive to one woman—her. By inhabiting otherwise unoccupied subject positions vis-a-vis male characters (or actors), such fans often conflict with others counter-hegemonically inclined to identify with strong, fictional heroines. Disputes between female "Scullyists" and "Mulderists" among *X-Files* fans, then, reflect such disparate and antagonistic mechanisms of identification (Scodari & Felder, 2000).

Soap multi-fan Gina was not as skeptical as other teen subjects in her devotion to the youth-oriented *Days of Our Lives*. Referring to a quadrangle story involving high school and college-aged characters that had dragged on indeterminately, she revealed: "I used to badmouthe this longest-running, overdone polygon, but then I realized that it keeps me glued like nothing else. DOOL [*Days of Our Lives*] fan for life, I guess." Such ardor, characteristic of romantic individualism as exhibited by McRobbie's (2000) pop star fans, can also be appreciated in the following introduction: "I'm an *ER* addict. I'm 15 years old from California USA and I can't get enough of this thing that has been in my life for so long."

Tangential Comments

Tangential posts, whether about personal or public issues, appear or develop from pertinent commentary on most Internet bulletin boards and are sometimes labeled with a prefix such as "TAN" or "OT" (off topic). Here, they ran the gamut from everyday life experiences, tastes, and activities, including interest in and/or perspectives on other TV series, to major news events. "Newbies," in fact, stress personal anecdotes when introducing themselves into a given forum. *X-Files* enthusiast Mandy recounted her sister's devotion to the anime series *Sailor Moon*:

> It's a Japanese cartoon about the Sailor Scouts (?) Who try to save the world from the Negaverse. . . . And every time they get upset or dreamy a little tear drop or heart, respectively, form on their faces. Personally I find the show annoying, but that's just me.

In the process of bemoaning the state of the Doug/Carol coupling on *ER*, Lynne took a recollective detour to *Star Trek: The Next Generation*: "The worst for me was when Counselor Troi and Worf got together. I *SO* wanted Troi and Riker to finally get together." Discussion of alternative TV texts operating on similar terrain as those featured in this study, like the pop star profiles in *Jackie* (McRobbie, 2000), functioned to circumscribe a specialized field of familiarity, lifestyle, consumption, and identification, often influenced by gender and/or age.

Young and the Restless fan Zoe was moved by a teen romance storyline to offer a twist on the generational divide and *Jackie's* tendency to instruct teen girls to accede to their parents' greater wisdom, assumed to be consistent with traditional sexual mores (McRobbie, 2000):

> Parents today DO NOT PARENT!!!! They don't give guidance to children. . . . I think the state of teens having sex with 67% of both male and female by the age of 18 is just revolting!!! That's how parents are doing w/their kids today.

Yet, Zoe speaks from a sense of unique identity as a discriminating teen girl more desirous of romantic fantasy than actual sexual experience.

Sandi often strayed from discussing *General Hospital*, in one instance sharing her memories of bookmobile visits to her Montessori school and her love of reading psychological thrillers, in another critiquing sex education in her high school, and in a third reacting to a tragedy in Colorado:

> As many of you know, I too attend high school. . . . Mine is a clean, average, middle class school. The biggest problems we are used to having is hallway fights that are cleaned up and forgotten the next day. Now what? How do we stop our school from becoming the next Columbine High?

Individual preoccupation, as might be expected, is evident in this commentary. Sandi assumes a "class-less, race-less sameness" among users that would make her comment instantly identifiable (McRobbie, 2000, p. 69). Neither the gun culture, violent media content, materialism, economic injustice, nor hegemonic masculinity is implicated. Sandi's ultimate question, while poignant, appears to be rhetorical.

Discoveries, Implications, and Prospects

As Interrogate the Internet (1996) has argued, "broader cultural contexts" and commercial factors can be expected to pattern online circumstances and discourses just as they influence the offline world (p. 127). This inquiry has demonstrated that teen girl cyberfans negotiate their identities primarily by proclaiming, nurturing, justifying, qualifying, and/or repudiating allegiances to, in this instance, television texts, actors, characters, stories, and/or creators. The fictional filter based on and through which subjective identification arises, and the multi-level, multi-layered aspects of negotiation diagnosed by Gledhill (1988), engender revealing insights into the psyches of teen girls. The process is facilitated by the existence of Internet forums centered around mass-produced narratives in which such affinities are announced, promulgated, and/or debated using methods and modes of online interaction, including asynchronous commentary and rebuttal punctuated by signature files.

McRobbie's (2000) subjectivity of romantic individualism, activated via hegemonic codes, underpins these propensities and permeates other categories of input, such as defenses of teens and tangential remarks about everyday life. The relative uniformity of this subjectivity bespeaks the larger culture's success in defining teen girls as a unique consumer group despite, as in this case, the diversity of texts under negotiation, user participation in these negotiations, and mechanisms of identification. The mechanisms witnessed here include: 1) identifying *with* a character in the text and/or the celebrity portraying her/him; 2) placing oneself within the text by filling a vacuum in relation to a beloved character; 3) maintaining a position outside the text as a relatively detached onlooker experiencing enjoyment; 4) adopting a position outside the text in order to second guess or supplant creators in their roles; 5) adopting a position outside the text to associate or dissociate oneself with the experience of other users. The same user, therefore, can adopt various positions with respect to the same text, depending upon the character/actor/creator, event/plot, and/or issue involved. To the extent that another user may vary in her/his negotiation in a given case, negotiation with other participants proceeds from negotiation with the text. Many such negotiations revolve around age and/or gender difference.

As Brumberg (2003) has counseled, the presence or absence of resistive competencies that might be collectively accessed should be noted in academic evaluation of teen girl discourse. Only the two acknowledgments of media culture's oppressive standards of appearance reflected oppositional promise in interrogating McRobbie's code of fashion and beauty.

Moreover, respondents in this study might not fairly represent the totality of teen girls participating in these forums, since the very fact that they were willing to post rather than just read and to declare and often defend their age group could set them apart from others too timid to do likewise. There is no telling how many adolescent girls are "lurkers" or read-only users on these forums, and whether their reticence stems in whole or in part from the adolescent girl silencing that Gilligan (1982), Chodorow (1978), and Pipher (1994) chronicled. An interest in programs and/or genres not exclusively or primarily targeted to teen girls preordains, in the subjects observed here, a certain distinction to begin with. Yet the majority of their articulations of identity, and even their adamant defenses of teens, tended to endorse romantic individualism and other hegemonic values.

Although established decades ago, McRobbie's (2000) codes of romance, fashion and beauty, personal and emotional life, and pop stars and music, as elements of romantic individualism, are upheld in the investigation's data. Also borne out is the idea that computer-mediated communication, including its mechanisms of net presence, appears to enhance and capacitate otherwise existing proclivities rather than impeding or redirecting them. Teen girls may be encouraged by online anonymity to assert ideas and opinions when they are not in a position to negatively impact their face-to-face relationships or be judged based on physical appearance, but this does not guarantee that the ideas they expound will be more likely than in unmediated contexts to challenge the status quo. Future research might compare and contrast the online utterances of adolescent girls that are not based on an intermediating text with those that are. Scholars interested in scrutinizing overtly political debate and discourse might explore whether and how the Internet could serve to intensify and mobilize such activity among those adolescent girls for whom the buds of political awareness have begun to open.

References

Agre, P. (1994). Net presence. *Computer-mediated communication magazine, 1*(4). Retrieved September 29, 2003, from http://www.December.com/cmc/mag/1994/aug/Presence.html

Barlow, L., & Krentz, J. A. (1995). Beneath the surface: The hidden codes of romance. In J.A. Krentz (Ed.), *Dangerous men and adventurous women of the romance* (pp. 15–29). Philadelphia: University of Pennsylvania Press.

Barnes, S. B. (2003). *Computer-mediated communication: Human-to-human interaction across the Internet.* Boston: Allyn & Bacon.

Baym, N. (2000). *Tune in, log on: Soaps, fandom, and online community.* Thousand Oaks, CA: Sage.

Brumberg, J. J. (1997). *The body project: An intimate history of American girls.* New York: Random House.

Brumberg, J. J. (2003, November 24). When girls talk: What it reveals about them and us. *The Chronicle Review*, B7. Retrieved September 26, 2003, from http://plsc.uark.edu/FW/when_girls-Talk.htm

Chodorow, N. (1978). *The reproduction of mothering.* Berkeley: University of California Press.

Consalvo, M., & Paasonen, S. (Eds.). (2002). *Women & everyday uses of the Internet: Agency and identity.* New York: Peter Lang.

Gilligan, C. (1982). *In a different voice.* Cambridge, MA: Harvard University Press.

Glaser, B. G., & Strauss, A. L. (1967). *The discovery of grounded theory: Strategies for qualitative research.* Chicago: Aldine.

Gledhill, C. (1988). Pleasurable negotiations. In E. D. Pribam (Ed.), *Female spectators: Looking at film and television* (pp. 64–89). London: Verso.

Haag, P. (2000). *Voices of a generation: Teenage girls report about their lives.* New York: Marlowe & Company.

Hayward, J. (1997). *Consuming pleasures: Active audiences and serial fictions from Dickens to soap opera.* Lexington, KY: University Press of Kentucky.

Interrogate the Internet (1996). Contradictions in cyberspace: Collective response. In R. Shields (Ed.), *Cultures of Internet: Virtual spaces, real histories, living bodies* (pp. 125–132). Thousand Oaks, CA: Sage.

MacKinnon, R.C. (1995). Searching for Leviathan in Usenet. In S. G. Jones (Ed.), *Cybersociety: Computer-mediated communication and community* (pp. 112–137). Thousand Oaks, CA: Sage.

Mazzarella, S. R., & Pecora, N. O. (Eds.). (1999). *Growing up girls: Popular culture and the construction of identity.* New York: Peter Lang.

McRobbie, A. (2000). *Feminism and youth culture* (2nd ed.). New York: Routledge.

Nichter, M. (2000). *Fat talk: What girls and their parents say about dieting.* Cambridge, MA: Harvard University Press.

Nochimson, M. (1992). *No end to her: Soap opera and the female subject.* Berkeley: University of California Press.

Oksman, V. (2002). "So I got it into my head that I should set up my own stable. . . ." Creating virtual stables on the Internet as girls' own computer culture. In M. Consalvo, & S. Paasonen (Eds), *Women & everyday uses of the Internet: Agency and identity* (pp. 191–210). New York: Peter Lang.

Pipher, M. (1994). *Reviving Ophelia: Saving the selves of adolescent girls.* New York: Putnam.

Radway, J. (1984). *Reading the romance: Women, patriarchy, and popular literature.* Chapel Hill: University of North Carolina Press.

Regan, K. (2000, August 10). Study: Women now online majority. *E-Commerce Times.* Retrieved September 15, 2003, from http://www.ecommercetimes.com/perl/story/3996.html

Scodari, C. (2003). Resistance re-examined: Gender, fan practices, and science fiction television. *Popular Communication, 1,* 111–130.

Scodari, C. (2004). *Serial monogamy: Soap opera, lifespan, and the gendered politics of fantasy.* Cresskill, NJ: Hampton Press.

Scodari, C., & Felder, J.L. (2000). Creating a pocket universe: "Shippers," fan fiction, and *The X-Files* online. *Communication Studies, 51,* 238–257.

Shade, L. (2002). *Gender & community in the social construction of the Internet.* New York: Peter Lang.

Stern, S. (2002). Virtually speaking: Girls' self-disclosure on the World Wide Web. *Women's Studies in Communication, 25,* 223–253.

Way, N. (1998). *Everyday courage: The lives and stories of urban teenagers.* New York: New York University Press.

7

What if the Lead Character Looks Like Me?

Girl fans of Shoujo Anime and Their Web Sites

Kimberly S. Gregson

Introduction

In Rye, New York in the summer of 2003, I attended *ShoujoCon*, an anime fan convention (Shoujocon, 2003). At the most basic level of description, anime is animation made in Japan. In fact, early anime imports into the United States were labeled "Japanimation." In recent years, it has gained increasing popularity in the United States, due in large part to its active and fervent fans who meet at local and regional conventions, in local clubs, in Internet chat rooms, and through email lists. In fact, the first anime-specific fan groups were started in the United States in the late 1970s (AnimeUSA, 2003). "More than any others, anime is an industry run by fans" (Yang, 1992, p. 47).

Being new to the world of anime myself, I did not realize that the *ShoujoCon* conference was anything other than a general anime convention. Until recently, the audience for anime—which typically includes battles, robots, and lots of action—had primarily been young and male, the "skateboard and surfing consumer, but also the computer user" (Atwood, 1995, p. 82). However, that changed in the 1980s with the introduction of series such as *Sailor Moon* and *Pokemon*. As one writer on anime observed:

If I learned anything from Otakon (one of the largest anime fan conventions), it is that the stereotypical white, male, socially inept fan is dead. The people I saw at the convention constituted a surprising mix of genders and ethnicities, mirroring the multicultural composition of America in general. (Ruh, 2003)

At *ShoujoCon* I quickly learned that all anime does not revolve around robots, explosions, battles, and schoolgirls in short-skirted sailor outfits. I also discovered that the anime fan base includes many girls and young women. They were at the convention to see *shoujo* anime—animation created by women for a female audience with a strong female lead and stories that focus on topics of interest to girls. The situations reflect girls' real-life experiences at home and school. The conference had three viewing rooms open for sixteen hours each day as well as a variety of panels where fans exchanged knowledge and met industry professionals. Attendees ranked their favorite *bishonen* (male) characters and talked about which Web sites were worth visiting. My experiences at the convention piqued my interest in two broad topics—how girls and young women relate to *shoujo* anime, and how girls participate in the online anime fan culture. Since the content of these *shoujo* series is different in many ways from other types of anime, and since generally, girls and young women are not as active in creating Web sites as are boys and young men (Pastore, 2000), I wondered whether girl fans of *shoujo* were typical fans in terms of their online behavior. In order to answer these questions, I analyzed common themes and content across 43 *shoujo* Web sites created by college-age or younger females, since they made up the bulk of the audience at the convention.[1] Informed by the work of Angela McRobbie, this chapter considers the manner in which girls focus their discussions of the characters and stories in their favorite *shoujo* series on male characters and romantic plots. In addition, John Fiske's concept of a fan economy/community is used to examine these Web pages, focusing specifically on girls' textual and enunciative productivity.

Anime and Manga in the United States

Most anime series originally were based on popular manga series. Manga is the Japanese word for comic books, which in Japan are cheap disposable publications with frequently published new issues. In Japan, both anime and manga are part of mainstream popular mass culture, while in the United States they occupy a growing, but still niche, cultural position. There are manga in many genres, for all age groups (including adults), and for each gender, as well as, for example, separate sports comics, office worker comics, even comics about mah jong. The graphics range from simplistic to intricately detailed, making use of cinema-like techniques to tell their story. The action does not stay contained in the small boxes on the page. Instead, it can spill out onto adjoining cells and even take up a whole page (Drazen, 2003). Many stories are more than ten volumes long; the *Prince of Tennis* series has more than a hundred volumes published so far. Popular manga artists are considered celebrities in Japan. For example, Osamu Tezuka, known as the "god of manga," used techniques from the movies he loved in his comics. His *Tetsuwan Atom* comic series was animated and is known in the United States as *Astro Boy* (Poitras, 2002).

It is from these manga that anime evolved and became popular. People are attracted to anime for the thematic complexity and variety (Napier, 2001). As one fan explains: "A good anime should have a great plot with good animation and beautifully crafted characters with human failings, superb mechas [robots], fan service, great soundtrack and a cute mascot" (Siew, 1999). In Japan the audience is primarily children and young adults, but in the United States the audience includes teenagers and adults, in part because of the sophisticated story lines and visuals. As the viewing audience has matured, so have the stories (Poitras, 2002). Much like television and film, anime spans many genres, offers stories for all ages, and features fast action, well-written dialogue, and amazing music and sound effects (Harbison & Ressner, 1999; Poitras, 2002; Yang, 1992).

The first animated television series in Japan, *Tetsuwan Atomu* or "Mighty Atom," about a boy robot created by a scientist to replace his dead son, was also the first anime series to reach American television, in 1964, where it was dubbed into English and called *Astro Boy*. In 1967 the Japanese television series *Mach Go Go Go* was popular in the United States, where it was titled *Speed Racer*. One American anime fan recalls the effect that *Speed Racer* had on him: "My love of, and obsession with, Japanese animation began with two words: *Speed Racer*. I can vividly remember watching the show as a child. Its combination of humor, action and, to my young mind, sometimes shocking violence couldn't be found anywhere else" (Brittingham, 1998, p. 21). *Robotech, Battle of the Planets* and *StarBlazers* were popular on American television in the late 1970s (Drazen, 2003; Poitras, 2002; Reeseman, 2002; Yang, 1992). Overall, viewers enjoy series that focus on "meaningful and entertaining stories" and have "characters [they] can fall in love with," as well as having "interesting, unusual plots that [they] can sometimes identify with [their] own [lives]" (Brickell, quoted in Dubois, 1999). Others watch anime simply because it is "different from the usual animation/cartoon that you see in the on TV" (Shinkami, 1998), or because it targets a wider audience than Disney animated features (FireoFire, 1998). Most fans list favorites from two or more genres of anime, suggesting that it is often the anime experience itself that draws in fans, rather than specific story types (Napier, 2001).

A new wave of anime became available in the United States in the late 1980s, targeting teenage and older audiences rather than young children. This dramatically increased the audience for anime and made the audience more representative of the wider population. American movie-goers watched the post-apocalyptic film *Akira* on the art-house circuit (Atwood, 1995). In 1991 MTV introduced a wider audience to anime when it ran the adult series *Aeon Flux* (Corliss, 1999). During the week of August 21, 1996, *Ghost in the Shell* was the first anime movie to top the *Billboard* home video charts (Goldstein, 1996*)*. *Blood: The Last Vampire* (released in 2000) sold 120,000 units in the first month it was available for home sales, making it the fastest-selling anime title at the time (Anime on DVD, 2001). In 2002, *Spirited Away* won the Oscar for best-animated feature, and in 2003, three anime movies were eligible to be nominated for that same Oscar—*Tokyo Godfathers*, *Millennium Actress*, as well as the newest *Pokemon* movie (Academy of Motion Picture Arts and Sciences, 2003).

The popularity of both anime and manga in the Unites States continues to grow as the mainstream media attention each has received has attracted a wider audience (Napier, 2001). A small group of companies publish manga in the United States—translated into English but in the original right to left, back to front layout demanded by fans. In 2004, publishers announced that they would make available more than 1,000 new volumes, including 450 by Tokyopop, one of the bigger manga publishers. DC and Marvel Comics are also making plans to produce manga series in the United States (ICV2.com, 2004). In 1996 Walt Disney Enterprises signed a deal with Studio Ghibli to distribute the studio's anime productions, including Miyazaki's *Spirited Away* and *Princess Mononoke*.[2]

For the most part the above are examples of *shonen* anime—series targeting boys and featuring action, adventure, explosions, big robots, and battles where lead characters have a specific goal they must reach by the end of the series. The stories show the process characters go through to reach their goals—training, journeys, quests—and the companions who join them on that journey along the way. Characters discover their inner strengths by overcoming obstacles thrown in the way of reaching their goals (Davis, 2003b).

A second type of anime series, *shoujo*, is marketed primarily to a female audience. *Shoujo* anime deals with questions asked by both boys and girls—"who am I, what kind of person do I want to be, what would I be willing to do for love" (Davis, 2003a, p. 2). The stories focus on the internal decision-making process that the lead female character goes through to answer those questions (Davis, 2003b). The creator of "Through Rose Colored Glasses," an *Utena* fan site,[3] claims that this was her favorite series because of the "interesting characters, arresting visuals, humor, drama, and disturbingly dysfunctional sibling relationships." The creator of a *Paradise Kiss* Web site comments that she instantly fell in love with the series, that the "story line really caught my attention too because it was so original. For some reason I found it somewhat easy to relate to as well since it was about fashion design and I had once wanted to be a Fashion Designer myself." The *shoujo* series deal with emotions and relationships between the boy and girl characters; plot is usually secondary to relationship development. "Almost nothing happened, but you certainly knew exactly how everyone felt about whatever it was that wasn't happening" (Levi, 1996, p. 9).

In 1995, *Sailor Moon* was the first *shoujo* anime to be officially licensed for distribution in the United States (Drazen, 2003). Unlike other anime series available in the United States at the time, *Sailor Moon* was marketed to teenage and grade-school girls. The stories in this series maintain a balance between romance and action, with themes of love, valor, and compassion.

> In *Sailor Moon* you had a monster-fighting action story, a cute-girl showcase, an epic saga of heroism, and a timeless romance all in one. That's the kind of something-for-everyone formula TV execs and movie producers consider themselves lucky to have run across . . . but for *shoujo* it's not all that unusual. (Davis, 2003a, p. 2)

More than 60,000 copies of the *shoujo* series *Vampire Princess Miyu* have been sold on DVD and VHS since it was released, making it one of the most popular anime series (Reeseman, 2002). It is the tale of a young girl vampire who never ages and who spends her days hunting demon-gods who torture humans. "She is delicately beautiful, but self-assured and in command of formidable powers. And her beauty has a perverse edge: she remains a vampire who offers humans immortality in exchange for the blood she requires" (Amazon.com, 2002). *Fruits Basket*, a popular *shoujo* series released in 2003 in the United States by FUNimation, tells the story of a young orphan girl who lives as housekeeper in a house owned by two male students from her school and their older cousin. The episodes deal with the deepening relationship between the boys and the heroine as she learns their family secret—they turn into animals from the signs of the Chinese Zodiac when touched by members of the opposite sex. Many of the episodes take place in the local school and include situations with which all viewers can identify, such as taking exams and dealing with other classmates ({Fruits Basket} Official Site, 2002). The stories in the popular *shoujo* series *Revolutionary Girl Utena*, distributed by Central Park Media, involve students competing in fencing duels to win the right to possess "the Rose Bride" and the behind-the-scenes machinations of a mysterious person known as "End of the World." "*Utena* delivers for girls what *Star Wars* did for boys: a never-ending series of adventures that one can imagine themselves in, whether daydreaming or playing in the backyard" (Amazon.com DVD: Revolutionary Girl Utena, n.d.). This series also takes place in a school; there are scenes in dormitories and classrooms, locations that are familiar to young adults everywhere.

Shoujo heroines typically behave in certain ways that Drazen (2003) labels "*shoujodo*," or the way of the teenage girl. The heroines are expected to be meek and kind—feminine, even while they are fighting to defend their friends or homes. In series with violence, the heroine can be involved in the action, but she goes out of her way not to actually hurt anyone. Unlike the villains they face, *shoujo* heroines are able to cry (Drazen, 2003). However, they are not the helpless princesses of Disney fairy tales. Instead, *shoujo* heroines take the lead in fighting the monsters, even rescuing the prince when necessary (Davis, 2003b). The heroine of *Fruits Basket* is very loyal to the boys in whose home she works; she struggles to protect them from the outside world by making sure others do not find out about the family curse. Utena, the heroine of *Revolutionary Girl Utena*, faces ridicule at school for dressing in a boy's school uniform and has to engage in constant struggles against the powerful Student Council who control the fencing duels and access to the Rose Bride.

The more I learned about *shoujo*, I was struck by its differences from traditional pop culture content produced for girls—pop music, movie fan magazines, and fashion/beauty magazines, for example, which focus on boys and romance. In creating media content for the young female audience, the media industries tend to highlight only one aspect of girl culture—girls' interest in relationships and romance. For example, in her classic study of the British teen magazine *Jackie*, Angela McRobbie (1991) described the "code of romance" (p. 94) found in the images, articles, and short stories of the magazine. These magazines emphasized fashion and beauty in articles, while the short stories emphasized romance, trying to make it look "important, serious, relevant" (p. 95). The focus was on romance, not sex; this was true for the boy characters as well as the girls. These stories helped girls learn how to act as they changed from children to adolescents but also promoted a stereotypical view of the role girls played in society, which included pursuing boys, mistrusting other girls as well as their own boyfriends, and looking pretty. Girls fought to keep other girls from taking their boyfriends. Yet, the overall impression in these stories was that romance, when you can get it, was fun and worth experiencing (McRobbie, 1991).

Given the differences between *shoujo* and traditional media content targeted to girls, I expected female fans of the genre to be drawn to it precisely *because of* these differences. Specifically, given the emphasis on strong female heroines in *shoujo*, I expected to find evidence on girls' Web sites that they identified with and/or admired these female lead characters. That, however, proved not to be the case, as these fans were more likely to talk about the male characters.

Relationships and Romantic Pop Culture Idols

Teenage girls commonly talk about boys and romance; typically these conversations take place in the privacy of a girl's bedroom with her friends among wall posters of current idols from pop culture (Wulff, 1995). They cut pictures and posters out of fan magazines and hang them on their bedroom walls. Karniol (2001) discusses this behavior as the adolescent girl's way of fostering discussion with her girlfriends as well as making public her interest in boys. Girls tend to pick pop culture stars from the movies or music for their idols because these stars are part of the "real, shared leisure culture in which all girls can participate" (McRobbie, 1991, p. 185). Idols can be selected from pop culture stars because they are convenient and safe—they expect nothing from the relationship since they are not really part of the relationship. Male pop stars, especially those with feminine features, can serve as practice love objects as young girls begin to think about being in a romantic relationship. Girls tend to be drawn to boys who are cuddly, like stuffed animals (Karniol, 2001). In fact, feminine features are attractive not just to adolescents; most adults also prefer male and female faces with more feminine features. These faces are considered more emotional, cooperative, and honest (Perrett et al., 1998).

There are many such male characters in *shoujo* anime series who are known collectively as *bishonen*, which "literally means 'beautiful' (*bi*) + 'boy' (*shonen*) . . . Generally, the term '*bishonen*' applies to male anime characters who are young (approx. 13–17 years old), very attractive and pretty, and even feminine to some degree (i.e. they have long hair, slender builds, might be gay, etc.)" ("What are *Bishonen*"). They show strong emotions, are romantic, and are the inspirations for many of the Web sites studied for this chapter. In fact, there are more than 360 entries in the *Bishonen* Underground Web ring. On the "Kuroi Tsubasa the Excellent page," for example, the girl who created the site describes why she likes a particular *bishonen* character, a demon-fighting angel named Access Time:

> What this section is really for is to describe what I like about the darling Access. . . . Let's start with his looks. First off, I love his big glassy eyes. They're so cute. Then his hair. It's purple for one thing, and it's long for another. . . . He started off his appearance with a low ponytail at his neck, but later has worn some other cool dos like a high ponytail. Access sports a stylish shiny purple gemstone on his forehead, and he has elf-like ears. I'm not too big on the ears myself, but they do add to his distinct image. Not to be overlooked, his single pointed tooth gives him a vampire-like bad boy appearance. . . . Access is also sensitive, we can see, when he gave Chiaki advice about not hurting Maron.

Romantic relationships between the boy and girl characters interest girl fans of *shoujo* anime series, and many choose to make Web sites about their favorite series. The "*Hana Yori Dango* Percolator" site describes the plot of that series as: "fun fun fun manga and anime (and movies) about a scrappy little lady named Makino Tsukushi and how she deals with the snobby kids at her elite high school. Oh yeah, and there's lurve and serious teen issues (well, do I really need to say this is relative?)." The creator of "Heart's Melody," a site about the series *Marmalade Boy*, describes why she created her site: "Eventually, as I became more entangled and interested in their [the main boy and girl characters] relationship, my feelings had to have an outlet . . . and voila, the idea of a site came to mind."

Everything on the "DandT" site revolves around the relationship between a boy and girl, Doumyouji and Tsukushi, characters in the series *Boys Before Flowers*. The site includes a list of reasons why this couple should have gotten together. The girl who created the "The Andre Grandie's Shrine" explains the attraction she felt for the *Rose of Versailles* series:

> *Berubara* (the Japanese name for the series) is absolutely the kind of anime that grabs your heart and changes your life forever. This is one of the reasons why I dedicate this page to Andre Grandier and Oscar Jarjayes [the main couple from the story, one of whom is a woman pretending to be a man so that she can serve as a soldier in the royal court]: two people who were destined to be together and who fought for what they believed in.

It is not just boy characters in general that make attractive idols for adolescent female viewers. In the romantic short stories studied by McRobbie (1991), the boys were not just the normal boy-next-door type; instead, they were nice and interested in romance—in other words, idealized. Girls liked how different the boys in the stories were from the real boys they knew from school (McRobbie, 1991). This seems to be true of the *shoujo* Web sites as well. One Web site serves as a shrine to the boy/girl couple Miki and Yuuhi of the *Marmalade Boys* series. The creator of the Yuuhi tribute page says that Yuuhi is the character from which she feels the biggest impact; she calls him a "pure boy" who

> loves everyone around him. . . . He's a normal, down to earth kind of guy and who wouldn't love someone who cooks like Yuuhi does. . . . And that is why I love Yuuhi so much . . . he has so much real depth, his relationship have ups and downs, his character is flawed and still beautiful, and he still keeps real . . . he becomes even more noble than he once was.

Just as with McRobbie's observations about the girl's magazine stories (1991), the girls and young women who create these Web sites are interested in romance, not sex. They reject the more adult (as in pornographic) anime, known in Japanese as *hentai* or *ecchi*. None of the series represented by these Web sites are in this category, although in the series *Revolutionary Girl Utena*, rape and incest are strongly hinted at in the onscreen action and explicitly discussed by fans. Many people who are not part of the fan culture associate anime with the explicit sex-and-violence titles that were the earliest imports and that are still a part of the anime fan world as seen in the description of activities at a Texas fan convention:

> In Dallas last month (at an anime convention), audiences wildly applauded short anime films made by their fellow fans, more packed with tentacled sex monsters, cute high school girls with magical powers, and the inevitable panty jokes than anything imported from Tokyo....

And in one corner, a tall shape, hooded in black, shuffles menacingly in the corner—the rapacious "Overfiend" of the *hentai* sex horror anime movie *Urotsukidoji*, his slimy tentacles and, well, other protruding things, all made out of squeaking, gaily colored party balloons. (Lewis, 1997)

The fans who created these Web sites do not want to be associated with that part, the most socially disvalued part, of the genre. The creator of "CrazyAnimeFreak [miss you more]" mentions several times that her site is *hentai*-free. The grade school girl who owns the "*Hamtaro* Kingdom" site is particularly grossed out by the thought of *hentai* anime—she says that it is "ewww" and that anyone who likes it is a "sad little monkey."

In addition to being concerned with the relationships between the characters in their favorite *shoujo* series, many mention a variety of other anime series they have watched, indicating that these girls are often fans of anime in general. Moreover, during the course of my research, it became clear that the girls manifest many of the same characteristics of fandom as described by Fiske (1989, 1992), in particular two aspects of his cultural economy of fandom—textual and enunciative productivity.

Fans as Textual Producers

Fans are different from general audience members. They have been defined as being obsessed with something such as a band or an actor (Hills, 2002), and as having "excessive enthusiasm" (Jenkins, 1992, p. 12) as well as having a sense of community based around the object of that fandom (Ehrenreich, Hess & Jacobs, 2003). That obsession is demonstrated by almost encyclopedic knowledge of the object of their fandom (Jenkins, 1992). A high level of knowledge, or perhaps even better, having access to unique, if arcane, knowledge, is evidence of credibility in the fan culture. American fans of anime sometimes refer to themselves as *otaku*, which in Japan has come to be a derisive term for a geek or a person totally obsessed with something—video games, television, computers (Levi, 1996)—but which in America is more likely to mean "diehard fans who ignore everything else because they love anime so much" (anime fan quoted in Napier, 2001, p. 241). However, anime fans have many other interests and are not as single minded as the quote makes out (Napier, 2001).

Fiske (1989, 1992) identifies textual production as one aspect of the cultural economy of fandom; it is a way to share knowledge with other fans. Fans write fiction stories and create works of art. They write about what they think will happen or what should happen on their favorite television program, drawing on their deep knowledge of episodes and their opinions about character motivation. More than just creating stories, they share their stories based on favorite episodes and characters with other fans. These fan-produced texts help define the fan community by marking off what fans like and do not like (Fiske, 1989, 1992). Anime fans are what Napier (2001) describes as "committed media fans" (p. 242). For example, groups of fans will create English subtitles for episodes of anime shows that are not yet legally distributed in this country (Newitz, 1994; Yang, 1992); these subtitled tapes have been credited with fueling the growing interest in anime (Ito, 2000). Unlike texts produced by the mass media, fan texts are not meant for the general audience but rather for other members of the fan community, who have background knowledge of, and appreciation for, the object of the fandom (Fiske, 1992).

In the mid-1990s, increasing access to the Internet and the World Wide Web changed some aspects of media fandom. Perhaps most importantly, it became easier to find information and other like-minded fans (Gwenllian-Jones, 2003). The Web allows for a multimedia presentation of self; adolescents make use of their own text and photos as well as files taken from other sites. Web page search engines make even niche publications accessible to a worldwide audience. Usenet newsgroup search engines, as provided by DejaVu and then Google, make it possible to read and post messages to thousands of ongoing discussions from the user's Web browser. Companies such as GeoCities and Tripod provide free disk space, so anyone can put up a Web page or a multi-page, multimedia Web site for that matter.

Fan Web sites are a type of textual production. Fans of bands, movies, and television series are all represented on the Internet, with discussion groups, chat rooms, and personal fan pages. Fans use the Web to share their story ideas and fictional writings. The Web makes worldwide distribution of these stories possible, not to mention the use of archives, multi-media, and links to related Web sites (Chandler-Olcott & Mahar, 2003). Fan Web sites provide episode background information and character profiles which are useful to people who do not have access to tapes of the series yet. Fan-written fiction can also make use of the episode summaries found on these sites as the foundation for the plots they develop (Chandler-Olcott & Mahar, 2003).

Anime fans have made extensive use of the Internet and the Web. While it might be possible in any given city to find hundreds of other fans of a particular movie star or a sports team, the anime fan base is smaller and more dispersed. There is a line in the mockumentary movie *Otaku no Video* (released in 1985 by Gainax Studios in Japan), where a new anime fan exclaims, "Why is it that you can be a sports fan and people won't mind, but like animation and EVERYONE HATES YOU?" (*Otaku no Video* cited in Yang, 1992). The Web makes it possible for anime fans to be part of the bigger and supportive online fan base, including the *shoujo* fan base. There is no shortage of Web sites created about *shoujo* anime series. For instance, "Amie's *Shoujo* Web Ring," a general Web site for sites about any *shoujo* series, had 17 active sites. The *Escaflowne* series fan listing had 19 sites included, and the fan list for the *Ceres* series had more than 100 entries. Constructing a Web site can be seen as a way to demonstrate technical ability and talent. Such sites enable creators to demonstrate a variety of types of knowledge, including knowledge about the series, technology, and Japanese language/culture as well as one's literary and artistic talent. Such knowledge accumulation is, according to Fiske (1989), characteristic of fandom.

Fans of any variety draw from many sources of information, one of which is the fan magazine with statistics and background information, pictures, and interviews. In the past, fans have consulted movie magazines and pop music magazines for information on the objects of their fandom. *New Type USA*, an English-language version of an anime fan magazine produced in Japan, includes information on anime series that have little chance of ever being licensed for distribution in the United States as well as information about what channels on Japanese television schedule anime. American fans can make little direct use of this information, but knowing it makes them a part of the anime fan community (Ruh, 2002). "In fact, this is part of what I find most appealing about anime, discovering 'new' shows or movies that have a following in Japan but haven't been released here yet" (Brittingham, 1998, p. 22). Fans accumulate knowledge about the programs, the actors, the producers and writers, which is then used to create new texts.

Textual production such as Web sites allows fans to share knowledge including knowledge about the anime series itself. These Web sites include episode synopses, character profiles, scripts, translations of text, and *scanalations*—images created from scanned *manga* pages with superimposed English translations. The creator of "*Emily's Hana Yori Dango Page! Ver. 2.0*" includes paragraphs of story summary interspersed with images from the program, and her opinions of the characters ("mostly interesting") and the music ("great"). "*Kyo's* Overview" page on the "Welcome to *Furuba* Magic! Site" includes story synopses, information on the voice actors, titles of songs on the soundtrack CD, and descriptions of the available merchandise for *Furuba* (the Japanese title for the *Fruits Basket* series). The character profiles are sometimes very detailed with information included from the DVD liner notes, but also from the manga comics, art books for the series, and official Web sites put up by the distributors.

A second type of knowledge included is of the Japanese language and of Japan itself. Girls on the "Vival La Diva" and "Dia-Chan's *Marmalade Boy* Page" sites mention studying in Japan in the near future. Levi (1996) points out that there is language snobbery among American anime fans; they prefer subtitled to dubbed versions, and they mix Japanese words into their everyday English. Using Japanese words plays up the *otherness* of anime and its exotic background but also the knowledge of the site creator (Napier, 2001). It is an example of how knowledge is capital in the fan economy; as in most economies, capital accumulation brings with it a measure of power and respect (Fiske, 1992). This is especially true if the person is willing to put their capital, in this case knowledge, on public display. Sites had Japanese words sprinkled throughout, sometimes with definitions for the uninitiated, while other times no translation or commentary was provided. Some examples include *omake* (extras or bonus), *arigato* (thanks), and *kawaii* (cute or adorable), which, according to Japanese teen magazine *CREA*, is "the most widely used, widely loved, habitual word in modern living Japanese" (Roach, 1999). "*Hamtaro* Kingdom" uses *kawaii* when describing the hamsters in *Hamtaro*. The "Energetic Heartbeats" site greets the visitor with the phrase "*Ohayou! Youkoso! ^_^*," which she translates to "Hello! Welcome!" Sometimes the Japanese text is not translated, as if to say, if the readers have to ask, they obviously are not part of the (anime fan) community. The "naishin [] innermost thoughts [] pure heart" site has the term *ja ne* as a menu option on the front page with no explanation. The "Andre Grandie's Shrine" site has the Japanese phrase "*Kimi wa kikari, boku wa kage!*" on the main page with no translation, and *sugoi* on the "about me" page. The site called "Dia-Chan's *Marmalade Boy* Page" includes a page of Japanese phrases that the site's creator is learning from a book on Japanese street slang. The "Energetic Heartbeats" site also has a lingo page with key Japanese words and definitions. The Japanese words are written with English characters and not Japanese symbols, which require special software to display.

There is a range of technical skill and talent, another type of knowledge, demonstrated on these sites. Some exhibit little evidence of technical talent, as the site creator appears to be making her first Web site, and the results are amateurish. These site creators, intrigued by the graphic and color possibilities, combine background images with foreground text in many colors and foreground images. "Grace's World" has a left-hand border of daisies and several sections of options down the very long page, each with their own graphics. The results are exuberant and colorful but difficult to navigate. These sites often consist of long pages, through which the reader has to scroll with no hint of what is to come. A general *bishonen* fan site entitled "*Bishonen* Guide ... understanding the pretty boys of anime, manga, j-rock, video games, and beyond" is one very long page, with a very long name, and with few links to other information or to other pages by the same girl. The creator of a general *shoujo* fan site named, plainly enough, "main" seems to realize how long her page is; she even puts text midway down the page to tell visitors to keep looking for the guestbook "PSSSST! The Guest books even *closer* now! Come on—you can make it! I have faith in ya!" On some sites, pictures are not optimized for use on the Web and take much longer to download than they should. "Dia-Chan's *Marmalade Boys* page" site has many such pictures—small on the screen but very large and slow to download.

At the other end of the technical-knowledge spectrum are those sites created by people with obvious training in Web design and layout. Their layouts are graphic intensive and some archive past layouts. These sites make use of cascading stylesheets, Javascript, and custom icons. The "CrazyAnimeFreak.cjb.net + Welcome to my crazy world!" site has a very professional look and archives a variety of past layouts, along with a collection of graphics visitors can copy onto their own site and an updated history on the front page. The main page of the "Autumn Crescendo" site for the *Marmalade Boy* series is a Photoshop image that has been divided up into many small graphics to create an image map; each individual image loads quickly, but the whole page is not viewable until every graphic has downloaded successfully.

Site creators often ask visitors to submit things for the site to help it grow. They rework the images, combining them with new music to make their own music videos, for example. There are fan-written stories and fan-drawn pictures. Some sites offer images visitors can download for their monitor background (wallpaper); they even offer thumbnail images so that image-gallery pages would load faster. "Kim's Anime Page" site has a thumbnail image gallery for the *Fushigi Yugi* series. "Jam & Jelly :: A *Marmalade Boy* Site" has wallpaper that the site's creator designed as well as several thumbnail-size pictures that visitors can click on to see images of Miki, a character from that show. The "Vision of *Escaflowne*" site has created *sprites* (small figures) for visitors to download. The "naishin [] innermost thoughts [] pure hearts" site has paper dolls (as well as a link to a site to download the software needed to play with them) and desktop icons. There are images, lyrics to songs, and multimedia pages from which visitors can download video and sound clips. With digital multimedia editing tools available on home computers, fans can make videos and add pictures to illustrate favorite songs.

A site called "Fire Starter—A Dilandau Albatou Shrine" has a multimedia section with separate pages for images from the *Escaflowne* series, wallpaper, music, and sounds. Examples of the literary and artistic talent are on display from both site creators as well as site visitors. Site creators are proud of the wallpaper graphics and the link banners they make for their sites. "The *Utena* Network" site has a variety of wallpaper graphics and WinAmp skins for the visitor to download. Others describe their efforts to scan images from manga comics and capture images from anime tapes (e.g., "DandT"). The creator of a *Fushigi Yuugi* site called "A Site Dedicated to Amiboshi" mentions that she does not own a scanner yet and so has to borrow pictures from other sites. At the other extreme, a site with the over-punctuated name "Yuuuuuuuuuuuuuuuuhi!!!!!!!!!! =) version 7 of EVENING SUN" has 138 images that represent a mix of those she scanned herself and those she has taken from other Web sites. "I scan these pictures for you guys, so please have the smallest respect back. If I didn't love my viewers, I wouldn't do this at all." She also has created a variety of wallpaper that she shares on her site for others to download. The "Welcome to *Hana Yori Dango* Magic!!" site has scans from a variety of series-related merchandise—the manga, poker cards, character art books, and fan-drawn comics. There are fan-written stories and fan-drawn pictures. For example, the "<- - - K y o t o m i s t 's Vision.of.*Escaflowne* - - -> ver 2.0" and "naishin [] innermost thoughts [] pure heart" sites both have sections for fan fiction and art.

One interesting type of information on these fan sites is the creators' reaction to spam, hackers, and people who copy their work. The creator of the "*Hamtaro Kingdom*" site is very worried about her site being hacked and her designs stolen. She wants to be friends with people who visit her site unless, as she mentioned specifically, the visitor is into hacking Web sites or is a "pervert." She threatens that her father, lawyers, and police will find them; then the hackers will have to pay money and go "to juvi"—assuming that the hackers will be other young people her age. Strangely though, she admits she had taken pictures from other sites without permission in the past. The creator of the "Forever Rain" site also asks people not to copy her pictures. "All content and information etc. were done and written by Cindy. Under no circumstances can you take this information. I have enough people copying me already. I don't like it one bit." The creator of the "Empty Movement" *Utena* fan site threatens anyone who borrows one of her layouts. "There are better designers to steal from, but should you be found out, I will hunt you down and kill you. Though it happens rarely, do NOT ask if I will make a layout for you. I won't." After all the threats and complaints about others borrowing the pictures and designs from their sites, many sites include disclaimers that they do not own the anime or the characters and ask that they not be sued.

Enunciative Productivity

Community is a big part of any fandom culture, including anime fandom. For some people, who may not have earned financial and cultural rewards in the general society, the praise of the fan community means a lot (Fiske, 1992). Fans are drawn to others who share their obsession, and getting together to talk about the object of their fandom is one aspect of Fiske's (1992) concept of enunciative productivity. Copying hairstyles and dress from that object are another part. Music fans dress up like the stars of their favorite rock bands (Fiske, 1989, 1992). They communicate with media creators about characters and plot. Fans take pride in the amount and variety of knowledge they master about the object of their fandom. A fan's stake in the community can be enhanced through learning and creating new things for the community. Perhaps most importantly, fans are happy to find they are not alone in their appreciation of the object of their fandom (Jenkins, 1992).

Writing before the widespread popularity of the Internet, Fiske (1992) considered enunciative productivity as a product of face-to-face communication. It is part of creating a social identity, and online communication can be part of that creation process. As seen in online guestbooks where visitors can both leave a message and read messages left by others, online communication seems to be another example of fan talk. The messages are part of an asynchronous multi-location conversation. For example, several pages identified by Chandler and Roberts-Young (1998) in their study of adolescents' Web pages were created by frequent chatroom participants; the creators of these sites referred other chatters to the Web site if they were interested. The software necessary to put a guestbook on a site is available free from a variety of sources so that even a technical novice can have a guestbook.

Many of the *shoujo* fan sites studied for this chapter link to a guestbook in several places throughout the site. Guestbooks serve as a way for site visitors to communicate with the site creator and subsequent visitors, who are encouraged to read all the public posts. With posts from countries as varied as Thailand, Singapore, Japan, France, and Canada, guestbook messages demonstrate the international nature of the anime fan community. The protocol for posts to the guestbook is to be positive, to gush over the work involved in creating the site, and to ask for more content. Posters may include a variety of different emoticons (smiley faces) in their comments. The *Boys over Flowers* site "Welcome to Leely.com" has many such messages, including this October 26, 2003 example: "Hello!! You have done a very good job on this Web site. I really like your translation. It's nice of you that you translate the new issue for us. I wish I had found you long before. Thanks!!" The "heart's melody" site for *Marmalade Boy* had almost 300 posts in her guestbook. One recent post said, "I love this shrine. It's very good considering you have a great amount of content whether it's information to images. Best part is, easy navigation (^_^) Keep up the good work." Positive messages can serve as validation for the site creator's efforts and, at the very least, provide a reason for the site's existence. Guestbooks are also a way for creators of other related sites to publicize their own efforts. Site creators sometimes comment about items in the guestbook. The creator of the "Kuroi Tsubasa The Excellent" site comments on some feedback on her site: "Ore-sama thanks you all for your adoration! I love you too! ^_^ Not quite as much as I love Fin-chan, but. . . anyways, a big thanks for the hits! I'm really glad that 'Kamikaze Kaitou Jeanne' is so popular, and that I'm so popular! How cool is that?"

One interesting bow to the fan community is the "spoiler warning." There is some tension between sharing the information one has about a series as a way of establishing credibility and not giving away a story ending to another fan. The creator of the "naishin [] innermost thoughts [] pure hearts" site warns of spoilers but explains their presence on the site by saying she "likes to be as comprehensive as possible on all her sites." On the *Escaflowne* site "Ashfae's Anime Page" is a warning that the pictures contain major spoilers. "So if you haven't seen all of the anime, you might want to think twice before looking around in here; I won't warn you again."

As a last observation about community, many of the sites appear dead and abandoned. For instance, "Grace's Café" has not been updated since January of 1999, the "Berusaiyu No Bara Presents: The Andre Grandie's Shrine" site since sometime in 1999, and the "HYD Project" and "Energetic Heartbeats" not since November 15, 2001. Site creators offer no explanation for abandoning their sites. There are no good-bye messages, no explanations for why the site creator left the site and its audience. The abandonment is exhibited in several ways. Most often there is a last updated message with a date several years old promising some format change or new content that never materialized. On other sites, pictures no longer load, links are no longer valid, and sub-pages disappear. One possible explanation is that the site creator began to like another anime series more than the one for which the original site was created or become a fan of something other than anime in general. Another explanation is that maintaining the site has become more of a time commitment than she could handle. This shows up several times in update messages—complaints about the pressures of school assignments, including promises to do more once the site creator graduates. Strangely, in some cases, even though a site has been abandoned, visitors continue to post messages in the guestbook complementing the hard work evident on the site. The disconnect between the site creator and the site visitors strains the idea discussed above about the role of fan Web sites as part of the fan community.

Conclusion

Fans often have a hard time identifying why they are drawn to any of the genres of anime. It may be the anime experience itself that draws in fans, rather than specific story types, since most fans list favorites from two or more genres of anime (Napier, 2001). The girl fans who made the Web sites included in this study liked a variety of anime series. They liked the stories, the graphics, and the action. The evidence suggests that fans of *shoujo* anime are much like other anime fans and fans in general. They like to demonstrate their knowledge of their favorite series. They like to participate in fan community activities as evidenced by the use of guestbooks and participation in fan listings and Web rings. They like to create pictures and write stories about their favorite series and characters and then to share them with others in the fan community.

Unexpectedly, girl fans of *shoujo* series do not necessarily identify with the female lead characters; they are not watching *shoujo* anime because they want to see talented girls or even kind, sweet girls become the heroines of the story. For the most part they show no identification with the girl characters but rather put themselves into the relationships she experiences. These girls use Web sites to talk about the boys they like from their favorite anime series and to put up pictures of those boys. This is very similar to the reasons girls were found to read romantic stories (McRobbie, 1991) and idolize pop stars (Karniol, 2001). McRobbie (1991) suggests that the media that target young women provide stories and images that young girls can use to try out aspects of their changing identity. She found that print magazines talked about romance and relationships between boys and girls. *Shoujo* anime series, which also target girls, provide material about relationships with idealized boys who are kind and more interested in romance than sex. If they were idolizing pop stars, many of these fan activities might take place in the privacy of girls' bedrooms because for those pop culture icons and relationships girls can find lots of other fans locally. For the anime fan, however, since the fan base is more widely dispersed, the Web makes it possible to find others who like the same series. The girl fans of *shoujo* anime have moved their bedroom discussion to the Web. The implications of moving what are typically private conversations into a relatively more public arena merit further investigation.

Even though, overall, girls tend to be online less and have fewer Web pages than do boys (Pastore, 2000, 2002), this chapter has shown that girls are active participants in the online anime fan culture. Moreover, the cultural economy of fandom in general is based on capital accumulation and textual production (Fiske, 1992), and the anime fan economy is no different. These girl fans exhibit a great deal of knowledge about the series, anime in general, and about making Web pages. They are rewarded for these displays of knowledge when visitors post positive comments in their guestbooks, engaging the site creator in fan talk about the series. They also are rewarded when other people contribute images and stories for the Web site—a form of capital exchange. The Web makes it easier to accumulate knowledge and to share texts among this dispersed fan base. Because of their relatively small numbers, this is especially true for girls who like *shoujo* anime. The Web provides an excellent arena for studying the cultural economy of specialized fan groups. More attention to the types of conversations about anime that take place on these Web sites may add to our understanding of the relationship between enunciative and textual production in what is, in all respects, a community built on text interactions.

Notes

1.   One difficulty with using Web sites as content is that there is no central listing of sites from which to sample. Many studies make use of an opportunistic sample of sites (e.g., Arnold & Miller, 1999). For this project, Web sites were selected using three criteria: 1) they had to be about *shoujo*

anime (or manga), and 2) they had to have been made by girls who were 3) less than 21 or still in college. Sites about *shoujo* anime series were located using Internet search engines, *shoujo* anime Web rings, and directories maintained by anime societies. However, as with personal home pages in general, many *shoujo* fan Web sites do not include identifying information other than perhaps a nickname used in the Web ring or fan listing. Each of these sites was then examined for information about the site's owner; if the site owner identified herself as a girl, then the page was put on the list to be examined further. Forty-three sites met all three criteria.

2. Additional background information on manga and anime can be found in works by Drazen (2003), Napier (2000), Poitras (2002), and Schodt (1996).

3. Web site names were copied from the source code of the first page of the site. They are reproduced inside double quote marks, with all of the spacing and punctuation exactly as the Web site creator intended.

References

Academy of Motion Picture Arts and Sciences. (2003, March 11). *Eleven films to compete for animated feature Oscar®*. Retrieved November 20, 2003, from http://www.oscars.org/press/pressreleases/ 2003/03.11.19.a.html

Amazon.com (2002). *Editorial reviews: Vampire Princess Miyu OAV (Vol 1)*. Retrieved December 1, 2003, from http://www.amazon.com/exec/obidos/tg/detail/-/B00005B8U6/102–3809994–8070521?v=glance

Amazon.com DVD: Revolutionary Girl Utena (n.d.). *Editorial Review: Revolutionary Girl Utena*. Retrieved February 5, 2004, from http://www.amazon.com/exec/obidos/tg/detail/-/B00000IBUK/qid=1076159965//ref=pd_ka_2/102–3809994–8070521?v=glance&s=dvd&n=507846

Anime on DVD (2001, July 1), *Manga [Entertainment, Inc.] celebrates 2001 with record breaking year*. Retrieved October 3, 2003, from http://www.animeondvd.com/news/pr.php?pr_view=159

AnimeUSA. (2003, September 22). *The Otaku unite at our nation's capital*. Retrieved December 1, 2003, from http://www.otakuunite.com/animeusa.html

Arnold, J., & Miller, H. (1999, March). Gender and web home pages. Draft of paper presented at CAL99 Virtuality in Education Conference, The Institute of Education, London Retrieved February 20, 2004, from http://ess.ntu.ac.uk/miller/cyberpsych/cal99.htm

Atwood, B. (1995, October 26). Japanimation rises to mainstream: Cartoons aren't just for kids anymore and anime isn't just for comic collectors. *Billboard*, 82.

Brittingham, D. (1998, June 22). A fan speaks: why I love anime. *Video Business, 18*, 21–22.

Chandler, D., & Roberts-Young, D. (1998). *The construction of identity in the personal homepages of adolescents*. Retrieved July 15, 2003, from http://www.aber.ac.uk/media/Documents/short/strasbourg.html

Chandler-Olcott, K., & Mahar, D. (2003, April). Adolescents; anime-inspired 'fanfictions': An exploration of multiliteracies. *Journal of Adolescent and Adult Literacy, 46*(7), 556–567.

Corliss, R. (1999, November 22). Amazing anime. *Time, 154*, 94–96.

Davis, J. (2003a, September). The rise of *shoujo* (and what it means). *Animerica, 11*, 2.

Davis, J. (2003b, September). Girl power. *Animerica, 11*, 28–29.

Drazen, P. (2003). *Anime explosion! The what? why? & wow! of Japanese animation*. Berkeley, CA: Stone Bridge Press.

Dubois, A. (1999). *Anime-The next dimension of animation.* Retrieved October 3, 2003, from http://www.phrozen-neon.com/apollo/

Ehrenreich, B., Hess, E., & Jacobs, G. (2003). Beatlemania: Girls just want to have fun. In W. Brooker, & D. Jermyn (Eds.), *The audience studies reader* (pp. 180–184). London: Routledge.

FireoFire. (1998, February 1). Re: Open question. . why do YOU like Anime? Message posted to news://rec.arts.anime.misc

Fiske, J. (1989). *Understanding popular culture.* Boston: Unwin Hyman.

Fiske, J. (1992). The cultural economy of fandom. In L. Lewis (Ed.), *The adoring audience: Fan culture and popular media* (pp. 30–49). London: Routledge.

{Fruits Basket} Official Site. (2002). Retrieved December 1, 2003, from http://www.fruits-basket.com/

Goldstein, S. (1996, October 5). Anime finds mainstream niche. (Japanese animation computer games). *Billboard, 108,* 71–72.

Gwenllian-Jones, S. (2003). Histories, fictions and *Xena: Warrior Princess.* In W. Brooker, & D. Jermyn (Eds.), *The audience studies reader* (pp. 185–191). London: Routledge.

Harbison, G., & Ressner, J. (1999). Amazing anime: Princess Mononoke and other wildly imaginative films prove that Japanese animation is more than just Pokemon. *Time, 154(21),* p. 94.

Hills, M. (2002). *Fan Cultures.* London: Routledge.

ICV2.com. (2004, May 18) *Over 1,000 Manga Volumes in 2004: 900+ from Top Three Publishers.* Retrieved online 4 August 2004, from http://www.icv2.com/articles/news/4927.html

Ito, R. (2000, January). The anime underground. *Los Angeles Magazine, 45,* 20.

Jenkins, H. (1992). *Textual poachers: Television fans & participatory culture.* London: Routledge.

Karniol, R. (2001). Adolescent females' idolization of male media stars as a transition into sexuality. *Sex Roles: A Journal of Research,* (January), 61–77.

Levi. A. (1996). *Samurai from outer space: Understanding Japanese animation.* Chicago: Open Court.

Lewis, D. (1997, July). Night of the Otaku. *Newsweek Japan.* Retrieved December 1, 2003, from http://www2.dgsys.com/~dlewis/anime1.html

McRobbie, A. (1991). *Feminism and youth culture.* Boston: Unwin Hyman.

Napier, S. (2001). *Anime: From Akira to Princess Mononoke: Experiencing contemporary Japanese animation.* New York: Palgrave.

Newitz, A. (1994, April). Anime Otaku: Japanese animation fans outside Japan. *Bad subjects: Political education for everyday life.* Retrieved December 1, 2003, from http://eserver.org/bs/13/Newitz.html

Pastore, M. (2000, September 13). *Teens not online as often as you think.* Retrieved October 15, 2003, from http://cyberatlas.internet.com/big_picture/demographics/article/0,,5901_459381,00.html

Pastore, M. (2002, January 25). *Internet key to communication among youth.* Retrieved October 15, 2003, from http://cyberatlas.internet.com/big_picture/demographics/article/0,,5901_961881,-00.html

Perrett, D. I., Lee, K. J., Penton-Vodak, I., Rowland, D., Yoshikawa, S., Burt, D. M., et al. (1998). Effects of sexual dimorphism on facial attractiveness. *Nature, 394,* 884–887.

Poitras, G. (2002). *Anime essentials: Everything a fan needs to know.* Berkeley, CA: Stone Bridge Press.

Reeseman, B. (2002, October 12). Home videos, TV, tie-ins encourage anime revolution. *Billboard, 114,* p. 40.

Roach, M. (1999, December). Cute, Inc. *Wired, 7.* Retrieved December 1, 2003, from http://www.wired.com/wired/archive/7.12/cute.html

Ruh, B. (2002, December 17). From here to Shinjuku: Toward a new type of culture. *popMATTERS.* Retrieved December 1, 2003, from http://www.popmatters.com/columns/ruh/021217.shtml

Ruh, B. (2003, September 4). From here to Shinjuku: A missive from the front lines of fandom. *popMATTERS*. Retrieved December 1, 2003, from http://www.popmatters.com/columns/ruh/030904.shtml

Schodt, F. (1996). *Dreamland Japan: Writings on modern manga*. Berkeley, CA: Stone Bridge Press.

Shinkami. (1998, February 1). Re: Open question. .why do YOU like Anime? Message posted to news://rec.arts.anime.misc

Shoujocon. (2003). *Chronicles*. Retrieved August 15, 2003, from http://shoujocon.com/chronicles/conronicles_whatisshoujo.html

Siew, S. (1999, September 13). RE: Could anime become "bigger" in the US than in Japan? Message posted to news://rec.arts.anime.misc

What are Bishonen and other questions. Retrieved February 5, 2004, from http://www.phenixsol.com/Miko/Misc/bishonen.html

Wulff, H. (1995). Inter-racial friendship: Consuming youth styles, ethnicity, and teenage femininity in South London. In V. Amit-Talai, & H. Wulff (Eds.), *Youth cultures: A cross-cultural perspective* (pp. 63–80). London: Routledge.

Yang, J. (1992, November 17). Anime rising. *The Village Voice, 37*, p. 46–47+.

8

Claiming a Space

The Cultural Economy of Teen Girl Fandom on the Web

Sharon R. Mazzarella

Introduction: Online Girls

An August 2000 study (Rickert & Sacharow, 2000) reported that girls between the ages of twelve to seventeen were considered to be the fastest growing group of Internet users. At the same time, however, while there is some overlap in activity with that of boys, girls use the Internet differently than do their male counterparts (Durham, 2001; Lenhart, Rainie, & Lewis, 2001; Tufte, 2003). As noted in the introduction to this book, this growing presence has not gone unnoticed, as general interest publishers, the news media, and the academy have all addressed various aspects of girls' presence online.[1]

In some cases, as mentioned in the introduction to this book, adults (including parents, the media, academics, and politicians) have expressed concern about the Internet as a force in the lives of young people. If it is not the Internet's potential to enable adults to prey on young girls that has generated moral panic, it is the fear that youth will be "seduced" by the technology itself and/or fall victim to the harmful messages contained therein. Indeed, as Newhagen and Rafaeli (1996) point out in their *Journal of Communication* "dialogue" on the topic:

> The popular press is already depicting the Net as having the power to snatch our babies right
> from the cradle and poison their minds. Based on virtually no empirical evidence, the

power to mesmerize and seduce our youth is being attributed to the Net, just as it was to television and film before. (p. 13)

When it comes to the Internet, Henry Jenkins argues that adults are so concerned with protecting our children from what they perceive to be potential harm, that they fail to take the time to understand what "our children are doing with media" (Jenkins, 1999a, ¶ 4). In his post-Columbine Congressional testimony, Jenkins identified three reasons behind adult (i.e., parental, media, and legislative) concern: 1) adults fear adolescents; 2) they also fear new technologies, primarily because they themselves lack the knowledge and expertise necessary to incorporate these technologies into their own lives; and 3) youth culture has become increasingly visible, making it harder than ever to ignore. "We are afraid of our children. We are afraid of their reactions to digital media. And we suddenly can't avoid either" (Jenkins, 1999b, ¶ 30).

Yet, while many in the academy, the media, and the general public have been busily wondering how to protect young people, in particular young girls, from the culture surrounding them, another group of scholars has turned their attention to studying the way girls negotiate that culture—in particular, the media. While acknowledging that the culture surrounding youth in general, and girls in particular, can often be difficult to navigate, more and more scholars have begun to focus on how girls actively negotiate this terrain. But it is not enough to understand how they, in their role as media consumers, negotiate the mass-produced, mediated messages targeted to them. New media, such as computers and the Internet, enable youth to be producers as well as consumers, and it is this newfound role as producer that warrants scholarly analysis. Indeed, Mary Celeste Kearney described this "emergence of girls as cultural producers" as "one of the most interesting transformations to have occurred in youth culture in the last two decades" (1998, p. 285). In calling upon youth scholars to move away from studying girls as simply consumers, Kearney reminded us that:

> If scholars involved in the field of girls' studies desire to keep current with the state of female youth and their cultural practices, we must expand the focus of our analyses to include not only texts produced for girls by the adult-run mainstream culture industries, but also those cultural artifacts created by girls. (1998, p. 286)

While there has been an increase in academic studies of girls and the Internet, Susannah Stern pointed out that most studies have focused on Web sites created *for* girls or other Internet content *about* girls, however, there has been little serious scholarly inquiry of sites created *by* girls. Stern's own work (1999, 2002a, 2002b) has paved new ground in this area. Examining personal home pages created by girls, Stern has shown how girls use the "safe" spaces of these pages as forms of "constructed self-presentations" (1999, p. 23). Of course, when provided with the opportunity to let their voices be heard, girls do not all say the same thing. In one study, Stern identified three different "tones" or "overall demeanour(s)" to girls' home pages: "spirited" (characterized by "excitement and optimism"), "sombre" (disillusioned, angry and introspective), and "self-conscious" ("unconfident and unsure," especially when debating how much to say online) (pp. 25–26). Given her findings in that study, Stern concluded: "Ultimately, it is clear that adolescent girls are speaking on the web—speaking in ways and words that are infrequently heard. . . . home pages provide girls with greater opportunity to openly express thoughts, interests, and to create a public identity" (1999, pp. 38).

In a study specifically examining girls' self-disclosure in their personal home pages, Stern found that such pages "do provide some adolescent girls with a unique and appealing place to speak virtually" (2002b, "Virtual" Voices on Girls' Home Pages section, ¶ 1)—offering girls the opportunity to talk about a range of topics from their childhoods to their emerging (and sometimes very active) sexuality. Indeed, in another study, Stern went on to specifically examine girls' sexual self-expression in their personal home pages. She found that by including poetry, short stories, biographies, photos, music and other features, girls' personal home pages enabled them to communicate widely "a sense of their developing selves—selves that inherently intersect with sexuality in adolescence" (Stern, 2002a, p. 283).

Girls and Fandom

Stern's research has documented that girls often include information about their favorite celebrities on their personal homepages (2002a, 2002b). In fact, she noted that pop culture artifacts such as celebrity photographs were the type of image most frequently included in girls' home pages "suggesting with what and whom girls wanted to be identified" (2002b, Reflecting on and Presenting Self Online section, ¶ 32). Moreover, Stern argued that girls' appropriation of celebrity photos, song lyrics, and even audio clips positions them as both users and producers of media content (2002a). "Indeed, in these sections of their home pages, girls' roles intermingle; a girl can produce entire portions of her site by appropriating media materials that simultaneously document her use and her understanding of them" (2002a, p. 279).

Some girls take this a step further by creating Web sites solely for the purpose of communicating their identification with (indeed, often their "love" for) one or another celebrity. Such sites typically focus on various youthful performers as Britney Spears, Justin Timberlake, Christina Aguilera, Ashton Kutcher, Hilary Duff, Chad Michael Murray, and so on. Heeding Kearney's call for studying "girls as cultural producers" (1998, p. 285), this chapter takes up the study of girls' fan sites as "cultural artifacts."

That girls would engage in such online expressions of fandom is not surprising. According to John Fiske (1992):

> Fandom is typically associated with cultural forms that the dominant value system denigrates—pop music, romance novels, comics, Hollywood mass-appeal stars. . . . It is thus associated with the cultural tastes of subordinated formations of the people, particularly with those disempowered by any combination of gender, age, class and race. (p. 30)

Other scholars, in fact, have noted the prevalence of fandom, particularly "romantic fandom" among teenage girls (McRobbie, 1991).

Given the historical tendency of adolescent girls to actively participate in fan culture, it is surprising there has not been more academic study of the phenomenon, in particular of the textual artifacts so blatantly linked with pre-adolescent/adolescent girls' celebrity crushes. One example of such artifacts are mass-produced teen idol magazines such as *Tiger Beat*, *Bop*, and *Twist*. Indeed, my female twentysomething-year-old college seniors and I share fond recollections of collecting such magazines in our younger years—of gently ripping out the promised glossy pictures and "giant centerfolds" of our celebrity crushes, and then lovingly taping them to our bedroom walls. Sure, in my time the pictures were of David Cassidy and Donny Osmond, while my students' "crushes" were Jonathan Taylor Thomas and Joey Lawrence. Still, the point is that, despite being a generation apart, we share a common cultural experience of adolescent girlhood. Indeed, the experience continues as these magazines still exist today. Surprisingly, however, little if any critical attention has been given to these magazines (Herrmann, 1998; Mazzarella, 1996; McRobbie, 1991) despite the growing body of scholarship on fandom, especially female fandom.[2] Where once girls participated in fan culture by ripping pictures out of magazines and posting them on their bedroom walls (Brown, Dykers, Steele, & White, 1994), they now scan pictures into their computers and post them on their own personal celebrity fan sites for the entire world to see.

Informed by Fiske (1992) and McRobbie (1991),[3] this chapter examines how girls participate in the culture of fandom using the relatively new medium of the Internet. Despite the fact that Fiske's theory of the "cultural economy of fandom" was penned over a decade ago, McRobbie's theory of "romantic individualism" some twenty-five years ago, and that neither mentioned the Internet, their ideas are startlingly relevant when applied specifically to girls' fan sites.

Focusing on one of the "hot," young, male actors of the early 2000s, Chad Michael Murray,[4] this qualitative content analysis examines the role of girls as cultural producers in the community of online fandom. Specifically, to quote McRobbie, I engage in "a systematic critique of [these sites] as a system of messages, a signifying system and a bearer of a certain ideology, an ideology which deals with the construction of teenage femininity" (1991, pp. 81–82). Yet, while McRobbie's original research examined mass-produced content targeted *to* teen girls, the present study examines content produced *by* teen girls themselves—content that results in girls' active role in this "construction of teenage femininity."

In her own research, Stern (1999, 2002a, 2002b) has acknowledged the difficulty of sampling from the entire universe of Web sites in part because "they may be modified, expanded, abridged, and even deleted from one moment to the next" (2002b, Method section, ¶ 1). That certainly is an issue for sampling from fan sites as well. An additional difficulty with this subcategory of Web sites is that the vast majority appear to be commercial sites—created by the corporately controlled, adult-run culture industries, and targeted *to* girls. One would have to wade through hundreds, even thousands of sites on a particular celebrity to find those created by fans themselves, and girl fans in particular. Stern's research (1999, 2002a, 2002b) has set a precedence for using a sample of 10 Web sites drawn from the larger universe of Web sites in the category under study. For this study, I typed "Chad Michael Murray" into the Google search engine, an act that yielded 174,000 sites, certainly not all of them fan sites, but nonetheless a daunting amount of material to wade through. Using a snowball sampling procedure,[5] I selected the first non-commercial, girl-created Chad fan site listed, "Chad Michael Murray: The Original British CMM Fansite" (CMM). I then selected the only one of her linked Chad sites that was non-commercial and girl-created, "Two to Tango" (TTT). Following this snowballing procedure, I was able to identify seven additional sites from the links provided by TTT, yielding a total of nine Web sites.[6]

Girls as Cultural Producers

While sharing many common elements in terms of what can best be described as *essential* content (pictures, bios, quizzes, filmographies, and so on), these sites evidenced a range of technical skills. One, "Daydream: A Chad Michael Murray Fansite" (DAY), was very sparse, primarily text; while another, Crazy for Chad (CFC), which was hosted by allstarz.org, was elaborate and flashy in its visual design, complex in its navigation, and dense in its content. The others were at every level in between. In some cases, girls clearly oversaw multiple Web sites, notably Thuy Le of Chad Michael Murray Online (CMMO), who included links to her twenty-five other sites! Some sites were updated regularly and recently such as CFC, TTT, CMM, and Bad for Chad (BFC) while others had not been updated for a year or two such as DAY, Fit Chad (FC), and Totally Devoted (TD). Take It from Here (TIFH), which had not been updated since 2002, included an opening graphic from webmistress Donna Mae warning: "due to my extreme laziness and busi-ness this site has been put on a brief hiatus . . . be right back." Still, I was able to access and navigate most of the site. Even if they have not been updated in years, these sites are still available to fans today, and so are part of the ongoing Chad fan community—a community in which, according to Fiske (1992, p. 30), "fans create a fan culture with its own systems of production and distribution." Fiske goes on to point out that fan activity within such a community has three primary charac-teristics: "productivity and participation" (37–42), "capital accumulation" (42–45), and "discrimination and distinction" (pp. 34–37).

Productivity and Participation

According to Fiske, "productivity and participation" refers to the act of fans creat-ing texts, which in Fiske's pre-Internet examples include such artifacts as zines.[7] Examples of fan productivity on these sites include the creation of Web sites them-selves, participating in online chat rooms, posting to a site's guest book, and con-tributing fan fiction.[8]

One type of productivity discussed by Fiske is "enunciative productivity" (1992, p. 37), an example of which is "fan talk"—i.e., the way fans talk to each other. This manifests itself on the Web through chat rooms and tag boards,[9] for example where fans can "converse" back and forth with other fans and in guest books where they can post messages. All of the web sites examined had tag boards and or links to other more "interactive," "conversational" opportunities for fans. While these will be dis-cussed at various points in the chapter, that discussion will center on their role in the structure of the fan site itself. In-depth analysis of the content of visitor post-ings is beyond the scope of this chapter which, instead, is to focus on girls as cul-tural producers—as creators of the Web sites themselves, an activity that fits Fiske's category of "textual productivity" (1992, p. 39). According to Fiske (1992, p. 39):

> Fans produce and circulate among themselves texts which are often crafted with production
> values as high as any in the official culture. . . . [Yet] fan culture makes no attempt to circu-
> late its texts outside its own community. They are "narrowcast," not broadcast texts.

Unlike zines, which often had a very limited distribution network, handed out to other fans at concerts or fan conventions, for example, the Web enables fans to cre-ate texts that are seen potentially by millions of people, many of whom are not a part of their fan culture. So, in the world of Internet fandom, Fiske's labeling of fan texts as "narrowcast" no longer holds. Indeed, as will be discussed later, postings in sites' guest books came from around the world. However, his observation that the qual-ity of fan-produced texts often rivals those produced in the official culture is upheld.

Many of the fan sites examined are flashy, complex, and visually stimulating. They are replete with photos (a staple of fandom) and are often divided into several pages such as biography, quizzes/polls, filmography, news/gossip, links to related sites, snail mail address at which to write Chad, and a guestbook. In fact, many sites contain the same photos, biographical information and other content. It appears that "sharing" across sites is common as the fan community seeks to support its members in whatever way necessary. In fact, webmistresses will often revise the design/content of their sites in response to reader comments, which are actively solicited. Many sites will specifically ask readers to tell what they think about the site.

In addition, given the community nature of online fandom, site creators will often apologize for things they feel are wrong with the site and/or for not updating regularly enough. For example, Jen of CMM wrote:[10]

> I've had several wonderful emails saying how much they enjoy this site. . I now it's not exactly top notch material but I don't really have the energy to perchase a dot com domain, I just wanted to create a simple yet good website about Chad for other Chad fans to gain info from and I think I've achived this.

In fact, a year earlier, Jen apologized to fans for not updating regularly and asked them to "forgive" her. Similarly, the DAY site began with an apology that "This site is under major construction, but it will be great—I promise!" Despite her opening promise that her site is updated "daily," Krys's most recent update of TD was nearly a month after her previous one, and it read: "Wow, it's been SOO SOO long since I've updated! What's with me?? Anyway I've been pretty busy judging for different award sites like the globes and such and also there's been NO chad NEWS!!! AT ALL!"

What is interesting about the production of Web sites is that, for teen/pre-teen girls, this represents one of the only opportunities they have to be creators of media content. It gives them power and the opportunity to flex their creative muscles. Yet, when given this opportunity, what are the underlying messages of the content they choose to produce? In many ways, these fan sites reproduce the content of the mass-produced teen idol teen celebrity magazines mentioned earlier. As documented both by McRobbie (1991) and Mazzarella (1996), mass-produced teen girl celebrity texts are characterized by—excessive use of visuals (celebrity pictures) and abundance of information (celebrity facts). According to McRobbie, the "profusion of 'facts' is accompanied with a profusion of visuals to the extent that the two seem to be integrally connected. The reader looks and knows and thereby acquires the necessary accoutrements of contemporary pop femininity" (1991, p. 171). McRobbie has described British celebrity-based "pop" magazines as displaying "a panorama of visuality focusing on male good looks" (1991, p. 154). While most[11] of the sites examined for this study offer dozens, sometimes hundreds, of pictures for users to look at and/or download, the most blatant example of McRobbie's point is found on the FC site. The importance of photos on this site becomes obvious immediately. The opening printed comments/updates are printed over a wallpaper collage of the same Chad photo repeated over and over. In addition, in these updates, the webmistresses, Kelly and Char, include information on new pictures in just about every update. "We have around 45 pics now, and we only choose the best Chad photos." Of course, the name of the fan site itself, Fit Chad, provides evidence of the importance of Chad Michael Murray's physical appearance. Again, however, this is not the only site stressing the importance of pictures. In evaluating the CMM site on her links page, Thuy Le of CMMO pointed out that CMM had "294 pictures to be exact," TIFH had "over 100 pictures," and another site (not included in this study) had "guaranteed eye candy!"

The sheer volume of pictures on each site necessitates that the webmistresses include smaller, thumbnail pictures that the visitor can click on to get a larger shot. Some of the sites explicitly divide up their pictures into categories such as pictures of Chad (as himself) and pictures of one or another of his characters, thereby enabling the visitor to narrow in on exactly the Chad pictures she wants to see. As McRobbie pointed out (1991, p. 171), such pictures "offer one of the few cultural spaces in which girls can stare unhindered and unembarrassed at pictures of boys. The endless supply of pictures including the pin-ups and the snapshots serve a sexual function based on looking." This follows Karniol's (2001) assertion that adolescent girls' fandom is a stage in their sexual maturation process.

But it is not just the pictures that contribute to this sexual evolution. As McRobbie (1991, p. 171) observed, teen idol magazines place a great deal of emphasis on information and on "facts" about the celebrities. In discussing the British pop music version of these magazines, McRobbie (1991, p. 169) noted:

> Any discussion of the music performed by these boys is almost wholly missing from the coverage given to them. There are no reviews, no comments on the kind of music they play, not even a description of their musical styles. Instead there is an overwhelming interest in personal information.

The same is true of the actors in the teen idol magazines examined by Mazzarella (1996). The majority of the information presented included birth dates, family information, favorite food, favorite color, height/weight, first date/kiss, upcoming movies/TV projects, and so on. Even when information related to their careers was presented, the questions asked were rarely about craft but were related more to personal feelings—what it was like to have their first screen kiss; how they felt about being away from home during the shoot; whether their character is at all like them. According to Mazzarella (1996) this enables the reader to feel as though she knows the celebrity on a personal level instead of as just some actor. As Mareike Herrmann (1998) pointed out based on her analysis of readers of the German teen girl magazine *BRAVO*:

> The girls had established illusionary relationships with their favorite [celebrity] boy by learning about their lives through the stories offered in the magazines. Gathering this information gives them the illusion of knowing the star, with whom they desire to have and imagine they do have an intimate connection. (p. 223)

The goal is to find out anything and everything about one's idol, even something as mundane as the fact that "Chad recently missed the NC State Fair because he has contracted bronchitis" (DAY). To that end, each of the nine sites examined included a section called something like "biography," "profile," or "quick stats."[12] In general the bio sections include such information as Chad's birthdate, eye color, zodiac sign, hair color, siblings, pets, and favorites (bands, foods, TV program, book, sports), and so on. Within the fan community, it is imperative for true fans to know the facts, and it is even more imperative for a site's webmistress to have the most up-to-date and accurate facts. Indeed, Jen of CMM took a lot of heat for having outdated information about Chad's dating status on her site. In her most recent posting/update, Jen wrote that "OF COURSE I was aware that Chad was no longer single as I own that 'YM' magazine where Chad confesses. . . ." But, as she went on to say, she just had not had time to update the site and asked readers to "please be patient" with her. In other words, within the fan community, it is acceptable to not have time to update, but it is not acceptable to *not know* key information about one's celebrity.

Even if girls do not create their own Web sites, there are other opportunities to participate in the community of online fandom. Site owners invite others to actively participate in the process by sending in their drawings, fan fiction, pictures, and so forth. As Kelly and Char (FC) wrote: "Ok, this is the fun part. All you Chad-fans can send in anything Chad-related to show your love for him. . You know, in case he sees this page for some cool reason." They then specifically invite contributions in the categories of fan art, fan fiction, poems, "fan-graphics" ("computer generated funky chad pics"), and "C-H-A-D" ("what do you think his name stands for?"). As Jen (CMM) wrote: "Anybody is welcome, providing that you are a massive fan who's reliable!" Similarly, she made a point of reminding readers that "This is the fan's page." Donna, the webmistress of TIFH, informed visitors that hers was "a site to ALL the Chad fans and is also a site where all contributions/donations are accepted. Every single thing from fan fiction to icons and also added links. We need everyones help! You'll always be credited." (By contributions/donations, Donna is referring to content not money.) The webmistress of CFC had an extensive "fan" section identifying ways in which fans could get involved including the Chad Birthday Project ("send in fanart and/or letters to Chad for his birthday"), Trivia (featuring the names of winners), and fan encounters (where fans could tell of their own encounters with Chad). In addition, the site offered a guestbook, tag board, and message board as well as a link to her Yahoo chat group.

Another way for girls to get involved is to sign the guest books of the sites they visit. In fact, the creators of various sites almost universally ask their visitors to sign their guest books, and often offer profuse advance thanks for doing so. As Stern noted, girls "seem eager for feedback and for a sense that the self they present on their home pages is both understood and appreciated" (2002a, p. 280). In this case, I am not sure it is a sense of self about which girls are seeking validation but rather an acknowledgment of their technical expertise and accumulation of Chad capital. For example, Thuy Le of CMMO wrote in her most recent update: "Version two was put on the net today. I love the new look and I hope you guys do too! Please leave your *positive* comments in the guest books" (emphasis mine). In fact, in her guest book's welcome message, Thuy Le instructs visitors: "DON'T WRITE ABOUT CHAD MICHAEL MURRAY because this is not his Official Website. Please leave your positive comments about this website ONLY." (A quick glance at the first two dozen or so postings reveals that almost all of the visitors ignored Thuy Le's directions and wrote about Chad, sometimes also mentioning the site itself.) Girls want and expect feedback on their sites. They specifically ask for visitors to tell them what they think of the site. Such praise for the site's creators is also linked with another of Fiske's characteristics of fandom: capital accumulation.

Capital Accumulation

As discussed by Fiske (1992) "capital accumulation" refers to the accumulation of "popular culture capital"—including both knowledge about the celebrity in question and actual *things* such as pictures and t-shirts. According to Fiske (1992, p. 42), "Fan cultural capital, like the official, lies in the appreciation and knowledge of texts, performers and events."

One way in which fans can gain popular culture capital is through knowledge acquisition. The fan/webmistress who is the first to break the latest "news" on the favored celebrity is afforded more status within the fan community. This is seen in the fact that many fan sites contain sections devoted to "news/gossip." The CFC site made a point of announcing that it contained links to 110 articles about Chad—links which the visitor could then click on to read the article of her/his choice. Even if the webmistress is not breaking the news, hers cannot be the only site not containing it. Similarly, most sites contain bios of the celebrity, typically featuring the same information as all the other sites. Again, yours cannot be the site without this "vital" information/knowledge. Within the fan community, such knowledge is power.

What brings the greatest power, however, is a real-life encounter with the celebrity in question. Although Jen (CMM) had never met Chad herself, she achieved a form of power and authority by association since she does "know someone who lives in NC where 'One tree hill' is filmed and she lives literally 30 feet away from his trailer. . no joke." Jen uses this connection to cite factual information allegedly gleaned from this friend. For example, Jen tells fans that, although Chad would love to answer all of their letters and emails personally, he has been too busy. "How do I know this? Good point." She then goes on to cite her friend, who found out from his publicity person "that he has been so busy with filming those two movies and starting OTH that he has not even had access to his mail and that he wants to read it personally."

Another form of capital accumulation I noticed on these Web sites was the accumulation of awards. Many sites, in particular CFC and BFC, proudly display numerous banners proclaiming awards they have won or include announcements of awards in their opening updates. Others urge readers to vote for that site in one ongoing contest or another. Christine, the outgoing webmistress of BFC, reminded visitors that "voting is still going on" for two specific awards, and she even went on to identify the specific categories in which she had been nominated—"Best Hosted, Best Layout, Best Webmast/Webmistress, and Best New Site." Verity, co-creator of TTT, wrote in her introduction: "If there is one thing I would love to receive from the making of this site it would be a official award, like from MenCelebs.com or Celebrity-link.com. I did actually get one from Celebrity Link for my previous Josh site!" Indeed, TTT, FC and other sites include links for visitors to click on in order to vote for their site in such contests. Of course, in this world of awards, the contests and votes all take place within the confines of the fan community. Being able to display an award-winning banner, however, affords the site's creator some measure of popular culture capital within that community and serves to further validate her technical expertise. As Fiske (1992) explains, such popular culture capital's "dividends lie in the pleasures and esteem of one's peers in a community of taste rather than those of one's social betters" (p. 34). In other words, according to Fiske, the popular culture capital one gains within the fan community does not necessarily translate outside of that community. He does, however, acknowledge one instance when this capital translates to the broader world—the "point where cultural and economic capital come together" (p. 43), for example, if fan collectibles can be sold for profit. In the case of Web site awards, it is conceivable that a girl could parlay her awards into an entry-level Web design job or even college admission.

Discrimination and Distinction

According to Fiske, "Fans discriminate fiercely: the boundaries between what falls within their fandom and what does not are sharply drawn" (p. 34). Basically, this phenomenon of "discrimination and distinction" implies that fans make value judgments about which celebrities are/are not worthy of fandom. Moreover, they distinguish between which fans are considered "true fans" and which are not. Despite finding ample evidence of the former in a pilot study examining fan sites for a range of teen idols (Mazzarella, 2002), none of the sites examined for *this* study took the opportunity to criticize or negatively evaluate another celebrity in order to build up Chad. There were, however, plentiful examples of the distinction between "true fans" and others.

One way this played out on these Chad fan sites was in the quizzes—a common feature in which readers could answer factual questions about Chad. For example, the TTT site featured a ten-question quiz that began:

> Think you are the biggest Chad Michael Murray fan in the world? Well, take this test to determine your knowledge on the guy and we will see if you are?

What happens when one, a mild-mannered college professor, for example, doesn't answer any of the questions correctly? In this instance, upon clicking on the box labeled "How did you do?" a new box appeared to inform me that I was "a disgrace to all Chad fans." Similarly, upon purposely getting only a couple of questions correct on a ten-question quiz on TIFH Web site, I was informed that I was "average," and "on the borderline of Chad knowledge." I then was advised to "review more Chad facts and try again later." Moments later, armed with my newfound Chad knowledge, I got almost all of the questions correct and was informed I was "AWESOME" and had "a high Chad IQ. [I] practically know everything on Chad. [I'll] rate high on finals and look sharp when [I] meet him." Interestingly, the TIFH site included three quizzes—one about Chad himself, and one each on two of his best-known TV characters. It is not enough to know about the celebrity, but the true fan should also know about his characters.

It is not the just the presence of quizzes that provide evidence of the discrimination and distinction of fans, but there also is a code of conduct appropriate for "true fans." The FC site began by announcing: "Hello, fans of Chad Michael Murray. First of all, we would like to point out that HE IS SO FIT. If you do not agree, then what are you doing on this site?!" One webmistress went on an irate tirade when she realized that postings to her guest book originally believed to be from Chad himself were determined to be the product of someone joking around. Jen, of the CMM site, wrote:

Okay, I've realised now that the signings in my guestbook that say they're by Chad are all friggin' posers. How could some stupid losers do that? I mean especially if they're obsessive Chad freaks, infact they're not worthy fans of his because doing something like that is so mean and in-considerate to other fans who may think it is really him.

Similarly, the webmistress of CFC offered a warning to "the very immature taggers" (i.e., posters to the site's tagboard) that:

the tagboard is a privilege. I don't appreciate one word posts that form sentences. I'm sick of having to delete posts like that and I don't think it's fair for me to take down the tagboard when other fans us it all the time, so my advice to those annoying posters–be mature and respect the tagboard and the taggers! Also keep in my mind that the tagboard is NOT a chat, therefore you can eliminate those ubiquitous questions like "Why isn't anyone talking to me?" Thank you for your cooperation.

While these examples serve to prescribe behavior appropriate for true fans, they also provide evidence of girls, in their roles as cultural producers, taking control of constructing the "culture" of their fan sites and, in some ways, of the Chad fan community as well.

Constructing Community

One thing that becomes clear is that the creators (and visitors?) of these sites are intentionally seeking to create (or join) a community. The most obvious example of this is the prevalence of links to other related sites and/or affiliates.[13] Webmistresses consistently ask visitors to link their own sites to the webmistresses's site and vice versa. In her research on personal homepages, Stern (2002a, p. 281) found links to be "present but not prevalent." In this study, however, links are quite prevalent and important. In fact, the CMMO site includes annotated links in which the webmistress, Thuy Le, describes the content of each linked site and then gives each a rating.

Certainly, the guest books play a role in constructing this community. As Jen of TTT wrote in her opening introduction to that site: "Ohh and sign the site guestbook! We want all the fans to connect and get more involved." In fact, that site includes a forum that others can join: "A paradise to all Chad Michael Murray and Joshua Jackson fanatics*~* This is the place for you to connect with other obsessers and share latest gossip!" Moreover, webmistresses are interested in the opinions of the site's visitors whether it be on the site itself or in response to some poll question related to Chad. For example, the TTT site was, at the time of data collection, running a poll asking visitors to identify their favorite of Chad's characters. The TIFH site had two questions, one of which was what the visitor would do if s/he saw Chad in a karaoke bar. Visitors to the CMMO site could select which female character or celebrity s/he would most like to see "romantically linked" with Chad. Poll takers could then click on a box to get the ongoing results of the poll displayed both in percentages and in the form of a bar chart.

Not only is a community created, but it is an international community. For example, Jen and Verity, the co-creators of TTT are from England and Northern Ireland respectively. The CMM and FC sites are British, and I also came across Chad sites from Sweden, Japan, and Italy.

Moreover, guest book postings on these sites are from visitors all over the world. As Thuy Le points out in the introduction to the "interaction" section of the CMMO site, that section includes "several communicational tools to use to interact with other fans around the world. Just simply click on the type that you would like to be a part of." Indeed, the phrase "be a part of" is very telling, as this is seen as girls' opportunity to commune with like-minded girls nearly everywhere. On this particular site visitors can do so through message board postings, a monthly poll, a quiz, and a chat room. Indeed, many of the sites examined include a section called "interactive"—meaning either/both interacting with the site itself (e.g., taking a quiz) and/or with other fans (e.g., message boards, chat rooms).

Constructing the Ideal Romantic Hero

Certainly, I, as researcher, selected the specific object of this online fan community—Chad Michael Murray. I could have picked any number of other male celebrities, many of whom model and embody a very different type of masculinity—for example, a more macho and/or aggressive masculinity. Despite my role in selecting the subject of this online fan community, it is obvious that the girls who also chose Chad as the object of their devotion did so for a certain reason and, in their roles as cultural producers, constructed him in a particular way—as a "nice" guy. Indeed, this correlates with points made by others studying adolescent girls and fandom (Herrmann, 1998; Karniol, 2001; Mazzarella, 1996, McRobbie 1991). From her analysis, Mazzarella (1996, p. 7) concluded that the male celebrities featured in U.S. teen girl fan magazines like *Tiger Beat*, *Bop* and *Teen Beat* were "'safe' romantic heroes—the kind of boys one could bring home to meet mom and dad." This tendency to promote "safe," "nice" romantic heroes to teenage girls is neither surprising nor unique to this genre. In her analysis of teen romance novels, for example, Christian-Smith (1990) found that readers "were repelled by teenage versions of the 'macho man' in books and everyday experience" (p. 108). Considering that readers of these magazines and creators of the Chad fan sites are roughly the same age as readers of teen romance novels, it can be assumed that they have the same preference in males.

Indeed, the creators of these sites are fond of pointing out what a nice guy Chad is. FC includes a transcript of an AOL online chat in which Chad, when asked which of his bad boy WB characters he most relates to, responded: "Probably neither. I don't think I identify with either more. I'm actually a nice guy." Similarly, Jen of CMM wrote:

> He goes home about twice a year, Christmas and the forth of July! He says he has inherited morals from his dad! His dad instilled such good morals in him he wakes up every morning and he just wants to do good things (aww!) He can't help it! He says it feels good. It makes him smile to do good things!!! Isn't that cute!

Not only does he like to do good things, but also these sites tell us that "he is very committed to keeping kids off drugs" (TIFH), "encourages kids to be drug-free (aww, bless)" (FC), "strongly believes in being drug free" (TTT), and is "an anti-drug role model for kids" (Day). In fact, the CFC site included a page titled "The Chad Dream: A Drug-Free Tomorrow" in which the webmistress included links to related sites such as Mothers Against Drunk Driving, relevant dates such as April as Alcohol Awareness Month as well as relevant song lyrics. In introducing the page, the webmistress wrote:

> Inso many interviews, we read about Chad wanting the young people of tomorrow to be drug-free. We can help Chad fulfill his dream by making more teens, and even adults more aware of the drug problem that is slowly, but surely, creeping into our lives.

Further evidence of his niceness relates to his taste in girls. As Mazzarella (1996) found with U.S. teen girl fan magazines, whenever a male celebrity was asked to describe his ideal girl, he never referred to physical appearance, preferring instead to mention things like sense of humor and personality. Unlike teen romance novels which rely heavily on a "code of beautification" in which the reader is made painfully aware that "beauty is the ticket to romantic success, power, and prestige" (Christian-Smith, 1990, p. 43), the physical appearance of the magazine reader is of absolutely no importance. There is no opportunity for her to compare herself physically to a fictional heroine or a "dream girl" and come up short. Given that girls themselves are the cultural producers behind these fan sites, it is not surprising that a similar phenomenon occurred on those sites that mentioned the kind of girl Chad likes. Typically, these sites say something to the effect that Chad's dream girl is "a girl who is sweet, supportive, with a great personality; a girl who can talk, and who can listen; can enjoy doing nothing sometimes; a smart girl" (TTT, FC). Adding to this are Chad's own words. In the 2002 AOL chat interview reprinted in FC, Chad responded to a question about the first thing he notices about a girl by saying: "Her eyes, and then personality."

Conclusion: Claiming a Space

Certainly, it would be tempting to dismiss the content of these Web sites as evidence that, when provided access to technology and the opportunity to be cultural producers, girls simply reproduce the content, style, and ideology of mass-marketed, commodified definitions of celebrity, romance, and fantasy found in teen idol magazines. Doing so, however, would be missing the more subtle significance of these Web sites. Yes, girls do recreate the conventions of teen idol magazines along with the latter's definition of an idealized romantic hero. However, closer inspection reveals that it is through the appropriation of these conventions and media content (such as photographs) that girls are creating a space for themselves—a space in which to engage in a practice (romantic fandom) that has been ridiculed, dismissed, and denigrated by the dominant adult culture for decades. Until now, this activity, like many elements of girl culture, has been engaged in in the privacy of one's bedroom and only in the company of one's inner circle of closest female friends.

As was discussed in the introduction to this book, many scholars see the Internet as providing girls with safe spaces—spaces where they can openly and freely discuss issues of concern to them (Furger, 1998; Hird, 2000; Stern, 1999). One such issue is their emerging sexuality (Stern, 2002a). While there was nothing overtly sexual on the fan sites examined for this chapter, Karniol (2001) points out that their adoration of male celebrities "represents a way for adolescent girls to cope with newly developing feelings at the point of transition to sexuality" (Discussion section, ¶ 11). Moreover, she argues that the tendency of girls to hang posters of various celebrities in their bedrooms is rooted in a "social function, serving as a focus of conversations with one's friends" (Discussion section ¶ 6).

In the case of these Web sites then, girls are taking control of their emerging romantic and sexual feelings. Moreover, by moving this fandom out of the bedroom and into cyber(safe)space, they are proudly announcing and celebrating their celebrity affiliation as well as, and more importantly, their developing romantic and sexual identities to an audience of potentially millions. Through the creation of an online fan community, the friendship circle is widened. Moreover, the fact that anyone can access these Web sites shows that these girls are no longer willing to hide either their fandom or their developing sexuality behind closed doors.

But it is more than claiming a space in which to celebrate one's romantic and sexual maturation that is evidenced by these sites. Girls are not only claiming a space, but also are constructing it to be a space in which they and other girls can feel safe. For example, the overall "niceness" of these sites, while in part an artifact of girls' construction of Chad as a nice guy, is further aided by the control the webmistresses exert over the content. FC announces: "Here you can leave a comment about Chad Michael Murray, or this site. We welcome any suggestions and/or outbursts of obssessive messages to Chad. However, please remember that small children may view this site, so please not too (cough)-y." Previously, they had warned viewers interested in submitting pictures of Chad not to submit naked pictures. Similarly, the creators of TTT warned visitors considering joining the Chad forum that they "do not tolerate any kind of harrasement in the group and have the right to ban members."

Girls are willing to exert their control as cultural producers to create the kind of atmosphere/environment they want and need. Paralleling their selection of a nice, safe, romantic hero to celebrate on their sites, these girls are also working to create a safe space for all who visit so that no one is offended by language or pictures that might be aggressive, vulgar or risqué. Considering that fandom, as Fiske (1992) notes, is typically denigrated by the mainstream culture, these webmistresses are actively constructing safe spaces for themselves and others to publicly engage in an activity they previously would have kept hidden. Indeed, a final component of this is the manner in which these sites celebrate girls, their fandom, and their roles as cultural producers. Through celebrating their fandom as well as their own textual productivity (and the productivity of others) in the form of positive guest book postings, fan art, fan fiction, awards, and links/affiliates, the girls who participate in this online fan community are claiming a space, both virtual and cultural, for themselves.

Notes

1. See, for example, Brown, 1999; Durham, 2001; Furger, 1998; Girls lured, 2003; Houston man, 2004; Jensen, 2003; Manning, 2001; Mitchell, Finkelhor & Wolak, 2001; Salzman, 1996; Sinclair, 1995; Stahl & Fritz, 2002; Stapinski, 1999; Stern, 1999, 2002a, 2002b; and Takayoshi, Huot, & Huot.

2. See, for example, Baym, 2000; Drucker & Cathcart, 1994; Ehrenreich, Hess & Jacobs, 1992; Harris & Alexander, 1998; Hills, 2002; Jenkins, 1992; Karniol, 2001; and Lewis, 1990, 1992.

3. McRobbie's original study of the British teen girl magazine *Jackie* was titled "*Jackie*: An Ideology of Adolescent Femininity," and was published as a CCCS stenciled paper in 1978. I am quoting from a version reprinted in a later book of collected essays by McRobbie (1991, pp. 81–134). In addition, in a later chapter of her book (1991, pp. 135–188), McRobbie analyzes 1980s issues of *Jackie* along with those of another British teen girl magazine, *Just Seventeen*. I refer to both studies throughout this chapter.

4. Chad Michael Murray is a twenty-three-year-old actor best known for his roles on the WB's teen-oriented dramas *Dawson's Creek*, *Gilmore Girls*, and *One Tree Hill*. With his scruffy, blonde hair, blue eyes, and little-boy charm, he fits the mold of the ideal romantic teen idol identified by Mazzarella (1996). When originally envisioning this chapter, I intended to focus on a more well-known young actor, Ashton Kutcher (perhaps best known at the time of this writing as Demi Moore's much younger boyfriend). Given Ashton's extremely high-profile and prevalence in the tabloids, there are thousands of Web sites (mostly commercial) devoted to him. I decided instead to focus on an actor who still appealed primarily to a pre-teen and teenaged female audience and who was less familiar to the tabloid-reading general public.

5. Snowball sampling is a type of non-probability sampling in which the researcher "initially contact(s) a few potential respondents and then asks them whether they know of anybody with the same characteristics" the researcher is looking for (Galloway, 1997). While I did not actually contact webmistresses for the names of others, I used their list of linked sites to find other related sites.

6. There were a total of fourteen sites linked from TTT, but not all were usable. Some were no longer accessible; one was more of an interactive list and not an actual Web site, and two were not in English. In addition to the two sites mentioned, the seven additional usable sites were "Take It from Here" (TIFH), "Fit Chad" (FC), "Daydream: A Chad Michael Murray Fansite" (DAY), "Totally Devoted" (TD), "Chad Michael Murray Online" (CMMO), "Crazy for Chad" (CFC), and "Bad for Chad" (BFC). The complete URL's for all five sites are contained in the References.

7. Zines are self-published documents on any number of topics including personal ramblings, music, television programs, politics, literary genres, etc. Some of the earliest zines were actually fanzines produced either by fan clubs or individual fans. In fact, the science fiction zine *The Comet*, published by a science fiction fan club, is generally considered to be the first fanzine (Duncombe, 1997).

8. Fan fiction is a common, long-standing feature of fandom in which fans write new stories using pre-existing media characters. For example, Krys of TD includes what she calls "Trory" fan fiction on her Web site. A Trory, as she defines it is "one that believes in Rory and Tristan"—two teenaged characters on *The Gilmore Girls* whom Krys and others believe should be romantically involved. Since the writers/producers of the program chose not to do more than hint at a mutual attraction between these two characters, fans took it upon themselves to write their own romantic stories using the characters and settings of the program.

9. According to one tag board service provider, a tag board is a "service that allows you to provide your visitors with real-time discussion without the complexity of forums" (Tag-Board.com). While this Web site spelled tag board as two words, many of the fan sites examined spelled it "tagboard." This chapter uses the spelling specific to the source being discussed.

10. To preserve the authenticity of the sites, I have left the girls' content exactly as written including spelling, grammatical, and punctuation errors.

11. A relatively recently created site, TTT did not yet have a photo gallery. As co-webmistress, Verity, apologized in the opening update for the site: "All the different sections are up for viewing except for the Multimedia section due to shear laziness on Jen and my behalf there isn't much there." Verity promises it will include not only a picture gallery but also audio and video clips and will be finished soon.

12. Krys of TD did not actually include the biographical details in her site. Rather, she had a page of links on which the visitor could click to visit other sites containing biographical information.

13. Thanks to my colleague Kim Gregson for explaining to me that "affiliates are people who put up a little graphic link that you create. It seem(s) to be a little stronger indication of approval and support than just a link" (Gregson, personal communication, 2004).

References

Bad for Chad (nd). Retrieved April 6, 2004, from http://bad-for-chad.cantdeny.com/updates.html

Baym, N. K. (2000). *Tune in, log on: Soaps, fandom, and online community*. Thousand Oaks, CA: Sage.

Brown, J. D., Dykers, C. R., Steele, J. R., & White, A. B. (1994). Teenage room culture: Where media and identities meet. *Communication Research, 21*, 813–827.

Brown, M. (1999). *Infogirl: A girl's guide to the Internet*. New York: Rosen Publishing.

Chad Michael Murray (nd). Retrieved April 3, 2004, from http://jens_creek.tripod.com/

Chad Michael Murray Online (nd). Retrieved April 6, 2004, from http://geocities.com/cmmonline/index.html

Christian-Smith, L. K. (1990). *Becoming a woman through romance*. New York: Routledge.

Crazy for Chad (nd). Retrieved April 5, 2004, from http://www.allstarz.org/~cmm/sitemap.html

Daydream: A Chad Michael Murray Fansite (nd). Retrieved April 3, 2004, from http://www.geocities.com/chad_murray_online/main.html

Drucker, S. J., & Cathcart, R. S. (Eds.) (1994). *American heroes in a media age*. Cresskill, NJ: Hampton Press.

Duncombe, S. (1997). *Notes from underground: Zines and the politics of alternative culture*. New York: Verso.

Durham, M. G. (2001). Adolescents, the Internet and the politics of gender: A feminist case analysis [Electronic version]. *Race, Gender & Class, 8*(4), 20–41.

Ehrenreich, B., Hess, E., & Jacobs, G. (1992). Beatlemania: Girls just want to have fun. In L. Lewis (Ed.), *Adoring audience: Fan culture and popular media* (pp. 84–106). New York: Routledge.

Fiske, J. (1992). The cultural economy of fandom. In. L. Lewis (Ed.), *Adoring audience: Fan culture and popular media* (pp. 30–49). New York: Routledge.

Fit Chad (nd). Retrieved April 3, 2004, from http://www.geocities.com/fit_chad/

Furger, R. (1998). *Does Jane computer? Preserving our daughters' place in the cyber revolution*. New York: Warner Books.

Galloway, A. (1997). *Sampling: A workbook by Allison Galloway*. Retrieved April 7, 2004, from http://www.tardis.ed.ac.uk/~kate/qmcweb/scont.htm

Girls lured via Internet [Electronic version]. (2003, September 27). *The Advertiser*, p. 42.

Harris, C., & Alexander, A. (1998). *Theorizing fandom: Fans, subcultures and identity*. Cresskill, NJ: Hampton Press.

Herrmann, M. (1998). "Feeling better" with *BRAVO*: German girls and their popular youth magazine. In S. A. Inness (Ed.), *Millennium girls: Today's girls around the world* (pp. 213–242). Lanham, MD: Rowman & Littlefield.

Hills, M. (2002). *Fan cultures*. New York: Routledge.

Hird, A. (2000). *Learning from cyber-savvy students: How Internet-age kids impact classroom teaching*. Herndon, VA: Stylus Publishing.

Houston man held in sexual assaults; He met 2 girls over Internet, police say [Electronic version]. (2004, February 7). *The Houston Chronicle*, p. 37.

Jenkins, H. (1992). *Textual poachers: Television fans & participatory culture*. New York: Routledge.

Jenkins, H. (1999a). *Congressional testimony on media violence*. Retrieved July 16, 2001, from http://media-in-transition.mit.edu/articles/dc.html

Jenkins, H. (1999b). *Professor Jenkins goes to Washington*. Retrieved July 16, 2001, from http://web.mit.edu/21fms/www/faculty/henry3/profjenkins.html

Jensen, L. (2003, October 18). Man, 25, is booked in Internet sex sting; He planned to prey on girl, 14, cops say [Electronic version]. *Times-Picayune*, p. 1.

Karniol, R. (2001). Adolescent females' idolization of male media stars as a transition into sexuality [Electronic version]. *Sex Roles: A Journal of Research*, (January).

Kearney, M.C. (1998). Producing girls. In S. A. Inness (Ed.), *Delinquents & debutantes: Twentieth-century American girls' cultures* (pp. 285–310). New York: New York University Press.

Lenhart, A., Rainie, L., & Lewis, O. (2001). Teenage life online: The rise of the instant-message generation and the Internet's impact on friendships and family relationships. Washington, DC: Pew

Internet and American Life Project. Retrieved March 17, 2004, from http://www.pewinternet.org/

Lewis, L. A. (1990). *Gender politics and MTV: Voicing the difference.* Philadelphia: Temple University Press.

Lewis, L. A. (Ed.) (1992). *Adoring audience: Fan culture and popular media.* New York: Routledge.

Manning, S. (2001). *Girl net: A girls' guide to the Internet and more!* New York: The Chicken House.

Mazzarella, S. R. (1996, May). Celebrity, romance, and fantasy: A critical reading of teen idol magazines. Paper presented at the meeting of the International Communication Association, Chicago, IL.

Mazzarella, S. R. (2002, November). The cultural economy of teen girl fandom on the Web. Paper presented at the meeting of the National Communication Association, New Orleans, LA.

McRobbie, A. (1991). *Feminism and youth culture: From Jackie to Just Seventeen.* Boston: Unwin Hyman.

Mitchell, K., Finkelhor, D., & Wolak, J. (2001). Risk factors for and impact of online sexual solicitation of youth [Electronic version]. *JAMA, 285,* 3011–3014.

Newhagen, J. E., & Rafaeli, S. (1996). Why communication researchers should study the Internet: A dialogue [Electronic version]. *Journal of Communication, 46*(1), 4–13.

Rickert, A., & Sacharow, A. (2000). *It's a woman's World Wide Web.* Media Matrix & Jupiter Communications. Retrieved July 7, 2003, from, http://pj.dowling.home.att.net/4300/women.pdf

Salzman, M. (1996). *Going to the Net: A girl's guide to cyberspace.* New York: Avon Books.

Sinclair, C. (1995). *Net Chick: A smart-girl guide to the wired world.* New York: Holtzbrinck Publishers.

Stahl, C., & Fritz, N. (2002). Internet safety: Adolescents' self-reports [Electronic version]. *Journal of Adolescent Health, 31* (1), 7–10.

Stapinski, H. (1999). What works—where the girls are: Women 18-to-34 congregate at the Girls on Web site to rant, rave, and review. *American demographics, 21*(1), 47–52.

Stern, S. R. (1999). Adolescent girls' expression on web home pages: Spirited, sombre and self-conscious sites. *Convergence: The Journal of Research into New Media Technologies, 5*(4), 22–41.

Stern, S. R. (2002a). Sexual selves on the World Wide Web: Adolescent girls' home pages as sites for sexual self-expression. In J. D. Brown, J. R. Steele, & K. Walsh-Childers (Eds.), *Sexual teens, sexual media: Investigating media's influence on adolescent sexuality* (pp. 265–285). Mahwah, NJ: Lawrence Erlbaum.

Stern, S. R. (2002b). Virtually speaking: Girls' self-disclosure on the WWW [Electronic version]. *Women's Studies in Communication, 25*(2), 223–252.

Tag-Board.com (nd). Retrieved on April 7, 2004, from http://www.tag-board.com/

Takayoshi, P., Huot, E., & Huot, M. (1999). No boys allowed: The World Wide Web as clubhouse for girls. *Computers and Composition, 16,* 89–106.

Take It from Here (nd). Retrieved April 3, 2004, from http://www.geocities.com/takeit_fromhere/

Totally Devoted (nd). Retrieved April 6, 2004, from http://www.geocities.com/plkrys14/totallydevoted

Tufte, B. (2003). Girls in the new media landscape [Electronic version]. *Nordicom Review, 24*(1), 71–78.

Two to Tango (nd). Retrieved April 3, 2004, from http://www.geocities.com/two2tango66/

9

Teen Crossings

Emerging Cyberpublics in India

Divya C. McMillin

Introduction

On October 29, 2002, six hundred excited children from rural Karnataka, a southern state in India, were bused to Bangalore, the state capital and information technology (IT) hub of the nation, for a computer lesson from India's President, Abdul Kalam. Sponsored by Samsung, the much-publicized event titled "The Students Internet World 2002" featured an hour of Internet instruction from Kalam during which students, most seeing a computer for the first time, learned how to access the Web and explore specific Web sites. In the president's closing comments, the computer emerged as a powerful symbol of modernity and the Internet a vehicle through which children could achieve their goals in a context of globalization (CIOL Bureau, 2002a).

India is an emerging giant in the IT industry. The implementation of the nation's aggressive economic liberalization policy in 1991 led to the sanctioning of Internet connectivity by the state-owned Videsh Sanchar Nigam Limited (VSNL)[1] in 1995 (Manzar and Ahmad, 2001). IT export revenues were estimated at $11 billion during 2002–03. Much of this business is geared toward the non-Indian consumer with as much as 66% targeted to the United States, 24% to Europe, and only around 8% and 0.8% to Asia and Africa, respectively (National Association of Software and Service Companies [NASSCOM], 2003). Much of IT's impetus depends on the endorsement of individual state governments. Of India's twenty-eight states, only nine (Andhra Pradesh, Karnataka, Tamil Nadu, Haryana, Madhya Pradesh, Gujarat, Punjab, Delhi, and Kerala) have developed reputations for fostering IT growth (Ahmad, 2001).

Since his inauguration as president in 2002, Kalam's recognition of the potential for IT growth in the country and specifically in cities such as New Delhi, Noida, and Gurgaon in the north; Bangalore, Hyderabad, Chennai, and Kochi in the south; Pune and Mumbai in the west; and Calcutta in the east, has driven the nation's aggressive efforts in IT development. In his opening speech at the 2002 Bangalore IT.com, Asia's largest IT and Telecom expo, Kalam said:

> Competitiveness is driven by knowledge power that is powered by technology, and . . . the latter is driven by resourceful investments. I will celebrate a day when IT goes to the remotest areas of the country, especially the seven states of the north-east so that they all could benefit from tele-medicine, tele-education, e-governance, e-commerce, and e-culture. (CIOL Bureau, 2002b, ¶ 3)

At the same event, Karnataka state Chief Minister S. M. Krishna, also well known for his strong support of IT development, presented the state as an experimental site for rural computer literacy. Union Minister for Urban Development and Poverty Alleviation Ananth Kumar introduced a mobile learning concept called *Vidya Vahini* (vehicle of learning) to take computer literacy to schools in remote corners of the state. Beginning with the district of Belgaum as a pilot site, twenty schools were fitted with the infrastructure necessary for computer classes (CIOL Bureau, 2002b).

National and regional rhetoric centering on the opportunities made possible by globalization and private investment in the Indian market reached a fevered pitch in 2003, presenting a critical point of entry for the study of Internet usage among Indian youth. This analysis examines computer usage among teen girls in India and is anchored in Bangalore, the capital of Karnataka state, well known for its high-tech environment and its significant population of software engineers, computer programmers, and English-speaking college graduates. Based on surveys and interviews conducted among students from five high schools and on participant observation at three cyber cafés across the city, this chapter attempts an understanding of the function of the computer and Internet for teenage Indian girls. Interviews and participant observation made it immediately apparent that email and Web surfing for the teen girls was a minor part of their leisure activity yet an integral component of a matrix of rituals of identity expression. As a new media technology, the computer, with television, facilitated a continuation of their gendered roles as residents of the private, domestic sphere. The role of television, therefore, will be highlighted as well in this study to provide a holistic view into how new media technologies reconfigure the gendered identities of the respondents and their membership in the patriarchal family structure.

Drawing from theory on national identity and popular culture, the analysis is situated within a critical framework that is characterized by its historicity and its attention to broad symptomatic and recurring issues. It recognizes that "the current, mediated age is neither singular, unified, nor stable . . . (and) places a premium on developing critical competencies that can look through transitional stages, technological conversions and cultural and identity boundary crossings" (Caldwell, 2000, p. 3).

Specific questions and considerations that guide the analysis are: What function does the Internet serve for teenage Indian girls? What conclusions may be drawn about the consuming agency of the girls in this study? What border-crossing communities, if any, emerge from the respondents' use of the Internet? A historicizing look at the Internet age in India, if it can be called such, requires an understanding of the nation's peripheral position in the world system of late capitalism. Information technology has transformed India into a necessary player in the global economy, yet much of its role involves providing support services for western nations.

Technology as a Vehicle of National Identity

Historicizing analyses of the Internet age in India squarely identify networks of technology as vehicles of centrist power, used by a hegemonic, patriarchal Indian state to discipline and rear obedient and progressive citizens (Sundaram, 2000). The ideological function of television in promoting upper-class, Hindu male supremacy and general female submission has been similarly theorized by media scholars (Mankekar, 1999; McMillin, 2003). Technology in India, ever since the first national government was constituted in 1950, has been touted as *the* essential factor that would lead the country to modernity and sustained development. Whereas the independence struggle was won through a resistance of technology[2] and specificity of place,[3] the first Indian government under Prime Minister Jawaharlal Nehru upheld technology[4] and dislocated icons[5] as markers of Indian national identity and progress. The Department of Telecommunications (DoT) and the national television network (Doordarshan) established in 1851[6] and 1959, respectively, grew to become monopolistic giants in this context. From the 1970s onwards a clear shift was evident in how national progress was construed. Rather than being anchored to physical industries such as steel, iron, and coal as developed by Nehru, the latter 1980s and the 1990s in particular, witnessed an emerging sense of the global, of India's role in the information age. Sundaram (2000) writes:

Development remained an issue but was reconstituted as a problem of *communication*. The way forward was computerization, networking, and a new visual regime based on a national television network. . . . The "national" was reaffirmed but through a new discourse which complicated the notion of borders and sovereignty that were so central to the old visual regime. (p. 276)

The complication of borders of patriarchy and sovereignty is seen in the phenomenal rise of private foreign and regional television networks in the early 1990s, providing stiff competition to Doordarshan metro stations in urban areas and presenting narratives of subversion and resistance (in a limited sense) to Doordarshan's nationalist ideologies (McMillin, 2001). The DoT was caught off guard with mobile phone connections rapidly replacing landline connections in the mid-1990s, allowing urban Indians to access the Internet, completely bypassing the cumbersome and inefficient methods offered by the DoT (Miller, 2001). Just as Doordarshan scrambled in the mid-1990s to provide more film-based, entertainment-oriented, and commercially-sponsored programming to meet the competition of private television (McMillin, 2001), the DoT, provider of local telephone services, is struggling to diversify to meet competition from private Internet service providers (ISPs). The long distance and International Data Dialing division of the DoT, the VSNL network, has now added thirty new cities to its ISP network. The government is aggressively privatizing this sector as it did television a decade earlier and almost immediately witnessed the birth of a giant, Reliance Industries, which developed a $3–$4 billion national fiber optic network to link 115 Indian cities by the end of 2002 (Miller, 2001). The year 2000 witnessed a boom in dotcoms with an average of three companies entering the scene each day. By the end of the year, there were an estimated 50,000 India-specific Web sites, many of which crashed soon after because of a lack of sophisticated customer-oriented business plans (Ahmad, 2001).

Internet connections and users within the country have grown from two thousand and ten thousand, respectively, in 1995, to over ten million and thirty million, respectively, in 2003. In a country with a population of over one billion, however, these figures are extremely modest. Overall, household penetration was estimated at 0.15% in March 2003 with 61% of that population subscribing to the Internet (NASSCOM, 2003). The domestic IT market is largely untapped (Rao, 2001), and Internet users are clustered in urban centers such as New Delhi (24%), Mumbai (22%), Bangalore (11%), Hyderabad (9%), Chennai (7%) and Pune (6%) (Amawate, 2003). Teenage students are estimated at just 8% of Internet users overall (Venkatesh, 2000). In the context of the privatization of hitherto state-owned industries such as television and telecommunication, audience and consumer identities rise in importance as compared to citizen and national identities. ISPs herald the idea of the global citizen leaping out of national borders tethered to the periphery of global development, to the postmodern centers of communication across the world, through the mere click of a button. Physical and virtual landscapes are changing as Bangalore reinvents its borders, making way for three new ring roads around the city, gleaming technological parks and gated housing developments on its outskirts, and KFC, Pizza Hut, and Benetton in its city centers. In this context, as Sundaram (2000) writes:

> (The) elite cyberpublic occupies a hybrid space which attempts to emancipate itself from the nation, its Border and its political public. The modes of representation allude to a fluid space where the nation is present yet thoroughly displaced, informed by a hybrid language, styles, and volatile mixture of both presence and absence. (p. 283)

As we shall see later, the teen Internet users in this study indeed occupied a fluid and hybrid space where national identity was not thoroughly displaced as Sundaram postulates but was heavily informed by the global (in terms of their consumption of foreign, English-language television, and such products as Gap and Levi's jeans or Pepsi and Coke products) and the local familial (in their participation in religious festivals, use of their native language at home, and conformity to conservative gender roles). Reminders of the nation through the centralized Doordarshan had diminished in the context of easy Internet access and satellite and cable television, yet national identity was reified through the teen girls' hybrid lived experience and daily rituals.

Locating the Popular

Cultural studies theory on popular culture provides a useful way to address the Internet habits of teen girls as rituals of identity representation. Such theory points to the emergence of urban teens as a marketing category as a product of late capitalism where varied commodities feed the adolescent need to assert individuality and difference from other developmental stages such as childhood or adulthood (Grossberg, 1992). As a working-class creation of yet another consumer market (Hebdige 1988), youth form a subculture that is " . . . a compromise solution between two contradictory needs: the need to create and express autonomy and difference from parents . . . and the need to maintain the parental identifications" (Cohen, 1972, as cited in Hebdige, 1979, p. 77).

In this quote, we see elements of the tension between global and local, where the teen stands poised between two overpowering attractions: the pull of the restless and hungry global market commodifying individuality and autonomy and the coddling arms of the local family where membership and participation in community rituals are just as crucial to identity formation. For the Indian teen girl, the patriarchal and conservative local is often a substantive force with non-negotiable stipulations on conduct within the home and school. Urban and rural girls are socialized into their adult roles and responsibilities at a very young age and learn, early on, that the public sphere is for the most part, a hostile space, and the domestic, private sphere is where they can and should excel (Verma and Sharma, 2003).

Global and local should not be seen as two extremes of the identity continuum, however, and as we shall discover later, teen girls constructed for themselves a realm of cultural identity that was dynamic, protean, and influenced by a variety of intermingling rituals: surfing the Internet or emailing friends, watching foreign and indigenous cable and satellite television, studying with friends for school exams, and participating integrally in religious festivals, to name a few.

Fiske (1989, p. 1) writes:

> Culture is the constant process of producing meanings from our social experience, and such meanings necessarily produce a social identity for the people involved. Making sense of anything involves making sense of the person who is the agent in the process; sense making dissolves differences between subject and object and constructs each in relation to the other. Within the production and circulation of these meanings lies pleasure.

The pursuit of pleasure for Indian teen girls is integrally connected with uncovering a safe space for that pursuit. Whereas teen boys may use public streets and playfields for outdoor sports and activities, teen girls are usually restricted to indoor activities (Shukla, 1994). Internet surfing at indoor locales such as cyber cafés, local libraries, or at home becomes a significant aspect of social experimentation. Hall (1980) underlines the importance of social and cultural contexts of meaning makers, contexts that are crucial in defining how meanings are constructed, and where the relationship between ideas and material forces is interrogated. The focus in this chapter on teen meaning makers necessitates a conversation on other media and social habits so that computer usage is appropriately situated within a matrix of technology usage within the household (see Bausinger, 1984; Lull, 1989). In the context of globalization, computer usage becomes an extension of other mediated activities, such as watching television or listening to music, that take teen users beyond immediate boundaries and contexts, transporting them from the classroom to the world described in history and geography textbooks and seen on television, from local family rituals to transnational spaces of youth identity.

Reading the Local

Descriptions of computer use in India have primarily been produced by industry research organizations such as Inomy and NASSCOM. Such research, while providing useful statistics on computer user demographics and patterns of usage, presents little to no information on the cultural questions of how, where, and why computers are used. To address such contextual questions, media ethnographers suggest varied methods of inquiry such as prolonged observation of family interactions around the television set (Gray, 1992; Morley, 1986; Seiter et al., 1989) or the computer (Giacquinta, Bauer, & Levin, 1993; Haddon, 1992); analysis of viewer's written responses on their enjoyment of a television program (Ang, 1985), and through interviews with consumers of a particular medium (Radway, 1991). Drawing from the methods provided by these ethnographers, this analysis includes participant observation and interviews at cyber cafés, phone interviews, surveys to obtain demographic information, and analysis of qualitative essays of the respondents.

Fieldwork was conducted during June, July, and August, 2003 in Bangalore, India. Around two hundred surveys addressing issues of computer, Internet, television, and cable usage were distributed among five high schools across Bangalore City. The high schools were chosen based on the profile of their students. High schools A and B were relatively affluent with students primarily from upper- and upper-middle-class families, while high schools C, D, and E targeted students from middle- and lower-middle-class families. The surveys included a demographic component where students responded to questions on name, age, gender, class grade, family income, and diversity of media available in the household. Sixty students (30%) returned the surveys and were then asked to write a qualitative essay on their computer and Internet usage, what they liked and disliked about the Internet, and what television programs and channels they liked most and least. For clarification and expansion of responses, follow-up phone interviews were conducted in August in Bangalore, and then in September and October 2003 and January 2004, from the United States. Cyber cafés provided a crucial site for the study of Internet usage and activity as well. Participant observation was conducted in three cyber cafés across Bangalore city. These were located on Brigade Road (a primary hub of youth activity characterized by shopping malls, pubs, and discos), in Indiranagar, a residential mixed-income area within the city, and in Whitefield, a residential semi-urban locality. This varied and staggered research design facilitated the collection of a wealth of quantitative and qualitative data and provided a rich look into the computer habits of the teen respondents in the study.

User Profile

Despite euphoria among national elites and IT entrepreneurs alike who consider India to have entered the Internet Age in 2000, at best, only 25% of the nation's population have access to the Internet (Ahmad, 2001). The results of the demographic component of this study reflected this dismal reality. Only 60 (30%) surveys were returned because a majority did not own computers or had never used one. Of those who had, most attended the more affluent Schools A and B at 29 (48%) and 25 (42%) in number, respectively. Four (6%) of the respondents were from School C and one each was from Schools D and E (2% each, respectively). All schools were private, selected for the greater likelihood that they would have computers than government schools. Private schools in India have historically exhibited greater responsiveness to market demands, better accountability to parents, greater religious and socioeconomic diversity in student bodies,[7] higher quality of facilities, and significant emphasis on English as a medium of instruction (Sudarshan, 2000). Karnataka is fourth highest among Indian states in the number of private schools that offer education up to the 12th-grade level (Department of Education, 1998).

The age of the students was distributed fairly evenly among fourteen (28%), fifteen (20%) and sixteen (22%) years of age, respectively, with 14% and 18% at thirteen and seventeen years, respectively. More students were in the ninth-grade (30%) than the eighth- (12%), tenth- (14%), eleventh- and twelfth-grades (22% each). More students (28%) reported monthly family incomes over Rs.25,000 ($556)[8,9] followed by incomes between Rs.10,001 and Rs.15,000 ($222–$333) and between Rs.20,001 and Rs.25,000 ($444–$556) at 14% each, respectively.

The Digital Divide

All the respondents had television sets and 94% had cable connections. Ninety percent of the respondents had radio sets, 58% had VCRs, 64% had stereos, 64% had computers, and 54% had VCD (video computer disc) players. As could be expected, those who did own computers were from higher-income brackets with as many as 43% of computer owners in the over-Rs.25,000 per month category, and 13% each in the Rs.10,001 to Rs.15,000 category, and the Rs.15,001 to Rs.20,000 ($333–$444) category, respectively. Also those with greater diversity of media technologies in the household were more likely to own a computer. Specifically, of the computer owners, 64% also owned a radio, television, and subscribed to cable; 62% owned a VCR, 75% a stereo, and 70% a VCD player. These observations correspond with those of Venkatesh's (2000) study of 1,000 urban families in eight Indian cities where a majority of computer owners were in the over-Rs.20,000 ($444) income bracket and were more likely to own a variety of media technologies and household appliances.

Those without computers (32%) regularly accessed email and the Internet at the neighborhood cyber cafés. At the time of the study, the average cost of a computer in India was Rs.30,000 ($667) and was most often made to order rather than pre-assembled. Internet access was relatively unproblematic with the low cost of computer usage at the plentiful cyber cafés. Forty-two percent reported that the price of computer usage per hour was between Rs.11 and Rs.30 ($0.24-$0.67) while 37% reported that the price for one hour of computer usage was under Rs.10 ($0.22). Overall, computer usage was not a major activity among the teens, with a majority (52%) reporting they used the computer under two hours each day.

Teenmail: Fast! Fun! Convenient!

From a survey of teen computer users in the United States, Tarpley (2001) notes that the Internet creates an increase in social activity in its ability to connect the user to distant acquaintances and a decrease in immediate family interactions. Email is the most popular online activity for children, and the "(i)nterpersonal lives and the computer activities of early adolescents reflexively amplified each other" (p. 553). In India, the typical Internet user is reported to be a male graduate or post-graduate student between nineteen and twenty-five years of age who surfs the Web for an average of 1.4 hours per day, spending as much as Rs.596 ($13.24) monthly on this service. Computer activity is mainly conducted at work, school, or in a cyber café, and the average user is not likely to own a computer. Cyber café visitors are again primarily male, of the same age bracket, and frequent these venues two or three times a week (Amawate, 2003). Venkatesh (2000) reports that email and Web surfing are primary activities of urban Indian computer owners and Internet subscribers, with these activities together comprising 99% of total online activity and around 47% and 43%, respectively, of total time spent on the computer.

These observations find resonance in this study where the interpersonal and computer activities of the teen respondents overlapped and amplified each other. Email constituted a primary online activity. The novelty of connecting almost instantaneously with friends and family in distant places had not yet worn off, and many communicated with classmates regularly as well. In particular, Internet surfing was a social activity where these teens browsed sites at home or in cyber cafés together for information on their favorite music and film idols. They were fascinated with the speed with which they could communicate with one another and with relatives in various parts of the country. Salma,[10] and Anusha, both fifteen-year-olds and Kavya, a fourteen-year-old, all ninth-graders from School A, said they spent as many as ten to fifteen hours on the computer per week, sending emails to cousins and friends all over the world. Forty-two percent of the respondents noted they used email most often with 18% and 16% stating they used email often and sometimes, respectively. Nikeeta, a fourteen-year-old ninth-grader from School A said she now rarely missed birthdays and anniversaries and regularly sent electronic cards to her friends and relatives. Umaz, of the same age, class, and school as Nikeeta, said she loved the fact that she could send and receive messages from her friends in seconds. Meghana, a seventeen-year-old twelfth-grader from School C wrote in her essay that the best part about using the computer was that she could "have contact with my dear ones even if they are very, very faa . . . a. .aar [*sic*] away from me."

As many as 64% had used a chat room at one time or another with 14% each using chat rooms often and sometimes, respectively. Many said that chat rooms were a good way to make friends. Naseema, a sixteen-year-old eleventh-grader from School B said chat rooms were a quick and easy way to keep in touch with people. Varsha, a sixteen-year-old of the same grade and school wrote that email and chat rooms were "Fun! Fast! Convenient for lazy people like me!"

Some others were enthusiastic about email but wary about chat rooms. Sameena, a seventeen-year-old from School B said that she used the computer:

> Mainly for browsing and emailing friends—just for chatting. I have a lot of friends, some from school, many from chat rooms. One of my neighbors has just moved to Dubai and I email her almost everyday. The chat rooms work well, but you know everything is not true—people can easily lie about who they are. You should really know what you are doing in a chat room.

Her friend and classmate Swetha, also a seventeen-year-old, added she stayed away from these spaces:

> I am not regular in chat rooms. I used to be when I was younger but now I know there's a lot of bullshit, lots of porn and cyber sex. You should be careful—many girls go in (to chat rooms), then cyber sex things happen, then they get really depressed.

These observations may sound inane to a Western reader for whom computers and Internet surfing are a common, everyday experience, so commonplace that computer usage among children is a well-researched area of study, reporting that 5.7% of Internet users are addicted to the Internet (Greenfield, 1999, cited in Tarpley, 2001). For the teen girls in this study, computer use was a small part of their daily activities with a majority of time spent on school, after-school classes,[11] and studying for examinations.

Observations at cyber cafés showed that although boys more than girls frequented these venues, teen girls were more likely than boys to use these spaces in groups. Small cubicles housed individual computers and two users could use a single computer at any given time. Group attendance at cyber cafés by female users stemmed from three reasons: first, because single females, teenage or adult, are still an anomaly in public spaces in the contemporary, yet conservative, urban Indian context; second, because the cyber cafés were themselves social spaces, playing popular Western and Indian music and providing inexpensive coffee and snacks; and third, because cyber café managers turned a blind eye to clients browsing pornographic sites. As Sameena, a seventeen-year-old from School B, commented, most of the time male users download pornography, making the cyber café a "rough place" for her to venture in alone.

Information at my Fingertips

Enthusiasm for the Internet as a source of information was common among all the teen respondents in this study. Many believed the Internet provided accurate and extensive information. Shruti, a fifteen-year-old tenth-grader from School B, found her homework moving much quicker with her computer. She said, "We can get various (research) materials without wasting our time looking for them in libraries." Chetana, a fourteen-year-old tenth-grader from School A, said "the best part of the Internet is that it provides right and useful information . . . and helps in improving my educational power." Sneha, a fourteen-year-old tenth-grader from School A said that information on any topic was available on the Web, making it very easy for her, as it was for her friends, Charitha, Vidya, and Pooja, also of the same age, grade, and school, to write class papers, find extra information, work on school homework and projects, and keep up with current news. Lisa, a sixteen-year-old eleventh-grader from School B said, "A world of information is at my fingertips," sharing her enthusiasm with seventeen-year-old Bhavana, a twelfth-grader at the same school who said she used the Internet for at least two hours a day because of the "unlimited information one gets on any topic under the sun."

Sameena, Swetha, and Samatha, all seventeen-year-old classmates from School B, said school assignments specifically required them to search for information online. School computers were available for email only from 9 A.M. to 10 A.M. and then from 3 P.M. to 4:30 P.M. with the rest of the day devoted only to browsing educational sites. Downloading pornography was strictly prohibited and closely monitored.

Tarpley (2001) notes that educators are concerned about the credibility of information on the Internet. While this concern may be extended to India, it is worth noting that education in Indian schools is highly standardized with state-wide examinations conducted at the seventh, tenth, and twelfth grade levels. Library and text-based research, hand-written homework assignments, and in-class examinations are given high priority so the reliance on the Internet for class materials is relatively limited. Also, the small sample of respondents qualifies the conclusions we may draw about the overt enthusiasm of the teen girls for the educational function of the Internet.

Untethered Communities?

Verma and Sharma (2003) rightly note that for teen girls in India, two-thirds of their leisure time is spent *indoors* whereas for boys, half of this time is spent *outdoors*. From their study of 100 adolescents and their families in urban India, these researchers found that girls spent most of their leisure time conversing with family members, helping out with household chores, watching television, talking on the phone, reading, or daydreaming. Only between 2% and 4% surfed the Web or emailed friends for an average of one to two hours a day.

The leisure rituals of the teen girls in this study echoed those of the adolescents in Verma and Sharma's group. For example, Samatha, a seventeen-year-old from School B said her typical day involved waking at 5:30 A.M., helping out with household chores and getting ready for school, attending school from 7:30 A.M. to 5:30 P.M., working on homework until 8:30 P.M., and then relaxing with her family and having dinner around 9:30 P.M., after which she was more than ready for bed. She hardly had time to watch television, let alone play on the computer. Computer usage was limited to homework assignments or email. Samatha's friend, Swetha also the same age and in the same school said:

> Leisure time for girls means different things depending on the family you are from. For most orthodox families, it is not good for a girl to be outdoors after 7 P.M. And most of us enjoy watching TV at home anyway.

Teen responses regarding questions on television revealed interesting patterns in their assertion of their fluid and overlapping global, local, and teen identities. First, they overwhelmingly preferred Hindi or English-language private television over national or vernacular language programming despite their South Indian backgrounds and knowledge of such native languages as Kannada, Tamil, Malayalam, and Telugu. More respondents (26%) enjoyed Hindi-language serials on the cable network Star Plus than on Doordarshan (4%), Sony Hindi (2%), or Zee TV (16%) because they focused on children, were comedies (such as *Shakha Lakha Boom Boom*, *Kichadi*, and *Khulja Sim Sim*), or were family oriented serials (such as *Kahani Ghar Ghar Ki*, and *Kyu Ki Sars Bhi Kabhi Bahu Thi*). Of the English channels, MTV, Channel [V], Cartoon Network and Star World were top choices. In particular, Star World serials such as *Xena Warrior Princess*, *Mad About You*, *Malcolm in the Middle*, and *Friends* were highly popular. Second, despite their reports of extensive television watching on the initial surveys, the respondents dismissed this activity as inconsequential, a common strategy among television audiences to downplay their susceptibility to television's ideologies and to identify with a dominant culture that views television as a waste of time (McMillin, 2002; Morley, 1986). Third, these teens were quick to deride Indian comedies as loud, brash, and slapstick. For example, Sunita, a fifteen-year-old ninth-grader from School A said:

> You know, I think none of the Hindi channels have figured out that TV is an intimate medium in your living room. They still treat it like it's the stage. There's a lot of shouting. Nobody talks softly into the mike. It's all (laughs), "Let's yell at the top of our lungs!"

With her friends Kavita and Leela, both fifteen-year-olds as well, Sunita discussed the lack of sophistication in production techniques. In particular, the girls discussed a horror serial on Zee TV, which they watched together regularly. Kavita said:

> The three of us were watching this program and instead of being scared, we were laughing at the camera angles. We were like, see if they had shot it from this angle (panning up with her hands) it could have been scarier. Like at a graveyard scene, they showed a hand coming up flat-on. Supposing they had shown this hand coming up (again demonstrating with her hands) like this, from the grave, it makes sense. It's small things like this that make us not want to watch Indian programs.

The deconstruction of Hindi serials as demonstrated in the exchange above shows that these respondents were sophisticated and critical consumers. Although sharply critical of Indian-language, particularly Hindi-language programming, they were ambiguous about English-language programming. Their top choices were MTV, Channel [V], the Discovery Channel, National Geographic, and TNT and, as with Indian-language programming, they coupled their viewing pleasure with criticism. Commenting on the variety of American serials on Star World and the proliferation of American products such as Coke and Pepsi on the Indian market, Sheela, a 16-year-old 10th-grader from School C said:

> It is worse than being colonized by Britain. Because at least then we retained at least a miniscule part of the Indian thing. Now we are being so vastly colonized by America, it's like, the way we talk, the way we dress, the way we eat, the way we think, *everything*—it's like this total hold on us.

Sheela and her friends, like Sunita and her companions, were nebulously positioned in shifting spaces of local and global identity. Sameena, a seventeen-year-old from School B, commented that much had changed in Bangalore since the late 1990s. There were many more out-of-state residents, particularly from the Indian northwest. There were also a lot more discotheques, not just in the city's downtown area such as MG Road or Brigade Road, but also in the fast developing suburbs along the new ring roads around the city. Hollywood, Speed Zone, Urban Net, Bunkers, and The Club to name a few discotheques, attracted midriff-baring teens in droves. Sameena said, "For orthodox parents who don't have much knowledge of the urban scene, who don't speak English very well, who are not computer savvy, their daughters are truly living in a different world." She continued that her social activities consisted of calling her friends on the telephone and hanging out at each other's homes or going out for lunch or coffee between classes or after school, unlike many other teen girls in her school who were slaves to discotheques and fashion channels on television such as Trend TV, Fashion TV and the fashion programs on MTV and Channel [V].[12]

The criticisms by the teen girls in this study were directed at uncritical fellow teens, the staid and pedagogical Doordarshan network, and the brash humor on serials on private and national television. Moreover, their awareness of consumerist ideologies on American television and advertisements was an expression of their autonomy and individuality. Through their derision, they stood apart from the local Indian consumer and simultaneously from the Western teens who, in their minds, were both passive and voracious consumers. Yet their positions were ambiguous in that they too were avid consumers of the global and local products in the market place. Most wore the more affordable indigenous imitations of Gap and Levi's jeans and T-shirts, and many also wore the more conservative and ethnic *churidhar*. For the teens in this study, criticisms of global and local were loosely mixed with consumption of the same. This ambiguity deserves further scrutiny and leads to a discussion of the function of the Internet in the lives of these teen girls, the nature of their consuming agency, and the emergence of border-crossing communities.

The Internet and Teen Crossings

Adolescence as a developmental stage is a relatively new concept that has gained importance in the context of modernity in India. Indian children are oriented to adult roles and responsibilities from a very early age. Adolescence as a transitional phase between childhood and adulthood is a product of urbanization, education, social change, and is in favor of upper-class boys. The teen girls in this study were from middle-, upper-middle-, and upper-class urban backgrounds and from English-speaking schools. They therefore occupied a privileged position where they could enjoy adolescence as a fairly well-defined phase in their lives. This is in contrast to their semi-urban and rural peers who are more likely to face pressures to get married or work outside the home to support the family (Verma & Sharma, 2003).

The awe expressed for the Internet and email by the teen girls in this study indicates not only the novelty of connecting instantaneously and relatively inexpensively to distant family and friends but also the excitement at entering a virtual public environment within a generally restricted context. Leisure time for these teens was limited to private spaces such as the home or indoor venues such as the library or cyber café, and the latter no doubt gained attraction for its ability (although limited) to "safely" connect them to a public space—a connection that did not defy societal norms of female propriety. The Internet then, although only a small part of leisure activity, was, like television watching, an important social one, allowing the girls to participate in a context and world beyond their local restrictive borders. The consuming agency of the teens in this study is an interesting dynamic. They derided the national network, dismissed brash Indian comedy, and criticized the colonizing forces of Western capitalism. Such a stance provided them empowering autonomous agency and set them apart from their imagined world of mindless fellow-Indian consumers. Yet it did not curtail their consumption of the same, be they televisual or market products. Such a dynamic points to the limits of critique in a capitalist environment where the market simultaneously provides a space for expression of identity through consumption of its products and also limits the expression of resistance by constructing alternatives (such as asceticism) as unrealistically radical options. The teens then were comfortable in their privileged position of being critics and consumers of the same products they criticized, straddling the best of both worlds. They also engaged in a certain level of critical response that allowed them to flit between nonchalance and avid consumption of the very products that were packaged in institutional and patriarchal high-risk rhetoric. In this manner, they could contest the norms of style imposed upon them (Hebdige, 1979) and express, within the constraints of Indian gendered norms for teen behavior, their sub-cultural identity of distinctiveness and difference from mainstream society (Barker, 2000) through new media (Boëthius, 1995).

Email and Web surfing then, for the teen girls in this study, although a small part of their leisure activity, was an integral component of a matrix of rituals of identity expression. As a new media technology, the computer, with television, facilitated a continuation of their gendered roles as residents of the private, domestic sphere. Yet, through their Internet and email connections and their consumption of limited global products despite the dangers associated with them, they could cautiously explore new boundaries in virtual space. While the nation around them lumbered along according to a postcolonial clock that measured only the nation's backwardness and developmental lag as compared to industrialized metropolies, the teen girls in this study were right alongside their Western counterparts, communicating through email, surfing the Internet, and watching current news, fashions, music albums, and comedies in concurrent time. With the increase in multinational corporations in Bangalore and more specifically, call centers, which are hungry for young, English-speaking, urban females fresh out of private, English-language high schools and colleges, the Indian teen may indeed be on the brink of discovering the Internet as a medium that transports her from private restrictions to public freedom. Of course, IT-based call centers are themselves hierarchical and may replicate exploitative colonial regimes, yet it is evident that the Internet, coupled with television, will present a formidable influence in how the Indian teenage girl visualizes her urban, gendered, and national identities and articulates her freedom and consuming agency.

Notes

1. The Videsh Sanchar Nigam Limited (VSNL), controlled by state-controlled Department of Telecommunications, provides International Data Dialling (IDD) services. VSNL links India to 237 countries, and has grown from providing one international line for every two thousand domestic lines to one for every nine hundred. Value added services (VAS) account for two percent of the total revenue. VSNL has invested in transmission equipment, and has established three earth stations at Calcutta, Madras, and Dehradun. VSNL also provides connections to the Internet and has one thousand subscribers with nodes in Bombay, Delhi, Calcutta, Madras, Pune, and Bangalore (Miller, 2001).

2. This was symbolically conveyed in Gandhi's use of the spinning wheel to make clothes and his preference for travel on foot instead of the railways.

3. This was represented through Gandhi's identification of the village as a space for anti-colonial nationalism.

4. National control was exercised through the central government's monopoly of the national television network, Doordarshan, and later through its establishment of a national computer network for public services.

5. Examples of Nehruvian modernity projects are dams and iron and steel factories.

6. Established as the Posts and Telegraphs Department in 1851 and commencing telegraph service in 1854 under William O'Shaughnessy, the Department of Telecommunications or DoT achieved its name only in 1984 when its telecommunications services were separated from its postal services.

7. Government schools enroll a majority of Hindu children while a majority of Christian and Muslim children enroll in private schools.

8. Indian currency is in Rupees (Rs.).

9. At the time of fieldwork, the exchange rate was Rs. 45 = $1.

10. Only first names are provided to protect the respondents' privacy.

11. Most students take after-school classes for additional coaching for class assignments and examinations.

12. Interestingly, with the rapid arrival of multinational corporations, global products, and foreign and Indian clones of American television programming genres and formats, School B enforced a strict dress code where female students could only wear the conservative *churidhar*—traditional North Indian attire commonly worn by urban women and girls all over the country and consisting of a loosely-fitting, knee-length dress and pants.

References

Ahmad, T. (2001). The state of Internet economy in India. In O. Manzar, M. Rao, & T. Ahmad (Eds.), *The Internet economy of India* (pp. 3–9). New Delhi, India: Inomy Media Pvt Ltd.

Amawate, V. (2003, October 1). Who is an Internet user in India? *Communication and Convergence Research*. Retrieved October 15, 2003, from http://www.ciol.com/content/idc_channel/idc_view/101032101.aspt

Ang, I. (1985). *Watching Dallas: Soap opera and the melodramatic imagination*. London & New York: Methuen.

Barker, C. (2000). *Cultural studies: Theory and practice*. Thousand Oaks, CA: Sage.

Bausinger, H. (1984). Media, technology and daily life. *Media, Culture and Society, 6*, 343–351.

Boëthius, U. (1995). Youth, the media and moral panics. In J. Fornäs, & G. Bolin (Eds.), *Youth culture in late modernity* (pp. 39–57). London: Sage.

Caldwell, J. T. (2000). Introduction: Theorizing the digital land rush. In J. T. Caldwell (Ed.), *Electronic media and technoculture* (pp. 1–31). New Jersey: Rutgers University Press.

CIOL Bureau. (2002a, October 28). Bridge the digital divide. Retrieved October 15, 2003, from http://www.ciol.com/content/news/repts/102102809.asp

CIOL Bureau. (2002b, October 29). Kalam asks industry to focus on IT. Retrieved October 15, 2003, from http://www.ciol.com/content/news/repts/102102902.asp

Department of Education. (1998). *Education in India 1992–93.* New Delhi, India: Government of India.

Fiske, J. (1989). *Reading the popular.* Boston: Unwin Hyman.

Giacquinta, J. B., Bauer, J. A., & Levin, J. E. (1993). *Beyond technology's promise: An examination of children's educational computing at home.* Cambridge & New York: Cambridge University Press.

Gray, A. (1992). *Video playtime: The gendering of a leisure technology.* London & New York: Routledge.

Grossberg, L. (1992). *We gotta get outta this place: Popular conservatism and postmodern culture.* New York: Routledge.

Haddon, L. (1992). Explaining ICT consumption: The case of the home computer. In R. Silverstone, & E. Hirsch (Eds.), *Consuming technologies: Media and information in domestic spaces* (pp. 82–96). London: Routledge.

Hall, S. (1980). Encoding/Decoding. In S. Hall, D. Hobson, A. Lowe, & P. Willis (Eds.), *Culture, media, language: Working papers in cultural studies* (pp. 128–138). London: Hutchinson.

Hebdige, D. (1979). *Subculture: The meaning of style.* New York; Routledge.

Hebdige, D. (1988). *Hiding in the light.* London: Comedia.

Lull, J. (1989). The family and television in world cultures. In J. Lull (Ed.), *World families watch television* (pp. 9–21). Thousand Oaks, CA: Sage.

Mankekar, P. (1999). *Screening culture, viewing politics.* Durham, NC: Duke University Press.

Manzar, O., & Ahmad, T. (2001). CEOs brave the internet carnage in India. In O. Manzar, M. Rao, & T. Ahmad (Eds.), *The Internet economy of India* (pp. 40–47). New Delhi, India: Inomy Media Pvt Ltd.

McMillin, D. C. (2001). Localizing the global: Television and hybrid programming in India. *International Journal of Cultural Studies, 4*(1), 45–68.

McMillin, D. C. (2002). Choosing commercial television's identities in India: A reception analysis. *Continuum: Journal of Media and Cultural Studies, 16*(1), 135–148.

McMillin, D. C. (2003). Marriages are made on television: Globalization and national identity in India. In S. Kumar, & L. Parks (Eds.), *Planet TV: A global television studies reader* (pp. 341–359). New York: New York University Press.

Miller, R. R. (2001). Leapfrogging? Indians information technology industry and the internet (Discussion Paper Number 42). Washington, D.C.: International Finance Corporation.

Morley, D. (1986). *Family television.* London: Comedia/Routledge.

National Association of Software and Service Companies. (2003). *The IT industry in India* (Strategic Review 2003). New Delhi, India: NASSCOM.

Radway, J. A. (1991). *Reading the romance: Women, patriarchy and popular literature.* Chapel Hill, NC: University of North Carolina Press.

Rao, M. (2001). E-governance services to unleash billion-dollar market in India. In O. Manzar, M. Rao, & T. Ahmad (Eds.), *The Internet economy of India* (pp. 48–59). New Delhi, India: Inomy Media Pvt Ltd.

Seiter, E., Borchers, H., Kreutzner, G., & Warth, E. (Eds.) (1989). *Remote control: Television, audiences and cultural power*. London: Routledge.

Shukla, M. (1994). India. In K. Hurrelmann (Ed.), *International handbook of adolescence* (pp. 191–206). Westport, CT: Greenwood.

Sudarshan, R. (2000). Educational status of girls and women: The emerging scenario. In R. Wazir, (Ed.), *The gender gap in basic education: NGOs as change agents* (pp. 38–79). New Delhi, India: Sage.

Sundaram, R. (2000). Beyond the nationalist panopticon: The experience of cyberpublics in India. In J. T. Caldwell (Ed.), *Electronic media and technoculture* (pp. 270–294). New Jersey: Rutgers University Press.

Tarpley, T. (2001). Children, the Internet, and other new technologies. In D. G. Singer & J. L. Singer (Eds.), *Handbook of children and the media* (pp. 547–556). Thousand Oaks, CA: Sage.

Venkatesh, A. (2000). Computer and new media technologies in Indian households: Based on a study of eight major cities in India. CRITO: Center for Research on Information Technology and Organizations. Retrieved November 25, 2003, from http://www.crito.uci.edu/noah/paper/IndianSurvey1.pdf

Verma, S., & Sharma, D. (2003). Cultural continuity amid social change: Adolescents' use of free time in India. In S. Verma, & R. Larson (Eds.), *Examining adolescent leisure time across cultures: Developmental opportunities and risks* (pp. 37–51). San Francisco: Wiley Periodicals.

10

"IM Me"

Identity Construction
and Gender Negotiation in
the World of Adolescent Girls
and Instant Messaging

Shayla Marie Thiel

Introduction

Each night in the two hours between the time thirteen-year-old Jordan finishes her homework and the time she goes to bed, her mother allows her to log on to the Internet and converse with her friends through an instantaneous textual communication medium called instant messaging or IM. IM is a technology free to anyone who has an Internet account and is able to download it. Through IM, Jordan (not her real name, as all names and log-in names have been changed for the purpose of privacy in this chapter) admits she has broken up with her first boyfriend, attempted to mend a hurting friendship, learned how to divide fractions, and spoken about many topics with dozens of people her age—most of whom she has met in person, but a handful of whom she has hardly spoken to even though they attend her middle school.

As she types furiously away in the abbreviations and slang that her online friends understand perfectly, it is clear she is a member of a distinct social network. In fact, she is just one of millions of teens who create and manage their own social worlds through IM. According to an extensive study conducted by the Pew Research Center on adolescent Internet use, close to thirteen million teenagers use IM—a number that is growing all the time—and say this technology-driven communication has a key place in many of their lives (Lenhart, Rainie, & Lewis, 2001). It is quickly becoming a preferred mode of communication for teens: 69 percent of those who use IM use it at least several times a week (Lenhart et al., 2001).

Adolescence is a time when people develop and construct identity and, notably, negotiate feelings of confusion as they straddle childhood and what most would consider some rather "adult" concerns (Erikson, 1950, 1968). This sense of confusion over identity suggests adolescence is a time of experimentation with different styles of communicating and articulating identity. In light of this, it is notable that 37% of the adolescent IM users say they have said something online that they would not have said in person (Lenhart et al., 2001). The implications of online conversation in contributing to adolescents' communication practices and articulation of self in relation to peers have yet to be realized, but the possibility that online communication may be changing these is worthy of study.

Adolescent girls are particularly prone to crises of identity as a result of dominant cultural and media discourses (Brown & Gilligan, 1992; Currie, 1999; Finders, 1996; Pecora & Mazzarella, 1999; McRobbie, 1982, 1997; Milkie, 1999; Orenstein, 1994; Pipher, 1995). They sometimes yearn for an impossible ideal of perfection in the eyes of their parents and peers (Brown & Gilligan, 1992), and develop both emotional and physical problems, such as eating disorders, as a result (Pipher, 1995). The media's world of adolescent girls is often characterized as particularly feminized—a world where physical beauty, sexual attractiveness, and product consumption (often that supplants both beauty and sexuality) supercede intelligence and creativity (Currie, 1999; Durham, 1996, 1999; McRobbie, 1982). These girls yearn for an unattainable perfection and niceness that is at odds with their desire to simply "be themselves," whether that may mean letting physical flaws or less-than-nice behaviors prevail (Brown & Gilligan, 1992; Simmons, 2002). Recent research also shows the culture of adolescent girls is rife with bullying among girls and that girls are ever prone to cattiness and cliquishness (Simmons, 2002; Talbot, 2002)—studies that may simplify somewhat the social world of girls but which foster an increasing need for further study of exclusionary practices among them. This research is important as ever, especially keeping in mind that gender discrimination throughout history placed women and girls at the margins of study and deemed them if not unworthy of serious study, then deviant or emotionally immature (Brown & Gilligan, 1992; Davies, 1993; de Beauvoir, 1949; Gilligan, 1982). This historical inattention to the study of girls' cultural practices has worked to effectively silence girls' voices or lump them together with the voices of boys, not taking into account the important differences produced by gender and the way gender works within culture (Davies, 1993; Orenstein, 1994; Pipher, 1995).

As society moves into the age of the Internet, technology becomes more important in girls' lives as a means to communicate (Fabos & Lewis, 2000; Lewis &

Finders, 2002) and articulate their own identities to the world (Chandler, 1998; Cherny & Weise, 1996; Stern, 1999, 2001). Computer-mediated communication also becomes a site of identity play and experimentation (Turkle, 1995). It has been interpreted as both a "safehaven" for open expression and normalized communication among genders (Evard, 1996; Turkle, 1995) and conversely, yet another place where voices are often silenced through patriarchal discourses (Cherny & Weise, 1996; Durham, 2001; Kendall, 1996).

Keeping in mind this dichotomy between IM as a safe space to experiment with identity and as a space where dominant patriarchal discourses also exist, I wish in this chapter to offer an intercession for understanding how adolescent girls, through their uses of IM, negotiate and articulate their identities, especially in regard to gender. Because IM allows for communication without body and actual voice, gender lines can be more easily blurred to allow for an intervention in the negotiation and construction of gender identity. Conversely, it seems that this identity is often played out in ways that are consistent with the real-life construction of "traditional" gender roles. The dichotomy can be understood by examining the conversations held through IM and listening to the voices of girls interviewed about IM's role in their lives.

Explicating Identity

Identity is a complex social construction created and sustained by a subject's location within a culture and society. Because of the sharp increase in computer-mediated communication and mediated technologies in general in the past decade or so, it becomes important to examine how identity construction has become increasingly complicated through its articulation in the uses of these communication technologies. Disembodiment, like that afforded through online communication, exists in a fast-paced, multi-networked environment among different races, genders, classes, religions, and across vast geographic locations, provoking myriad questions about how the lack of a body may shape and color one's perception of culture and one's location within culture.

It is at this juncture of cultural perception and cultural location that the concept of identity demands new attention. Ultimately there is no one, absolute identity because identity is bound up in cultural discourse or dominant notions of what it is to exist and behave in the context of society and culture (Foucault, 1974). From the moment of birth, humans mark themselves (and are marked by others) as they exist within cultural ideologies (such as the family, educational system, politics and government, religion, etc.). They cannot exist outside of the ideological systems—and discourses—that surround them because within these systems, humans believe they are in charge of how they behave, identify themselves, and function (Althusser,

1968). However, this is an illusion based on a person's understanding of the system; humans are born as subjects of these belief systems and are thus interpellated (or "hailed") by the ideology constantly surrounding them (Althusser, 1968, p. 245).

Judith Butler (1990) takes up the idea of hailing in her assertion that gender identity is a performed construction based upon dominant cultural discourses. Females and males "perform" what they interpret their gender to be based upon, what culture has taught them is the correct (heterosexual) interpretation of gender—for instance, that girls are "nice" and wear pink and boys are "boisterous" and wear blue. Stuart Hall (1996) furthers our understanding of the ever-shifting nature of identity, pronouncing it "never unified or coherent"—except when people wish to "construct a comforting story" or what Hall calls a "narrative of the self" as a way to feel some coherence about who they are and how they exist within culture (Hall, 1996, p. 277). Identity constantly shifts throughout a person's life and throughout daily experiences and, as Hall (1996) has argued, is in fact, fragmented, which hearkens back to Althusser's idea that fragmentation is a product of the Subject being hailed by the ideological systems within which she exists.

In the disembodied world of the Internet, identity is complicated through the notion of representation. Although many studies have discussed online representation in terms of falsehoods based on people intentionally misleading one another (regarding their race, gender, class, and other markers of identity), others have discussed identity in terms of "play" and experimentation through behavior, conversational, and textual manipulation (Stephenson, 1988; Turkle, 1995). Prevalent in nearly all research published on the topic is the representation of gender through online communication.

The study of identity within a computer-mediated environment offers a unique means to facilitate and grasp the fluidity of gender. In a 1993 issue of the *New Yorker*, an often-cited cartoon shows two dogs sitting at a computer, with one dog saying to the other: "No one knows if you're a dog in cyberspace." Indeed, no one knows whether persons are male, female, or a combination of both, unless they choose to signal gender to them through discourse—and even then, this may be an elaborate performance of gender or a play with identity that has little to do with the body a person owns in real life.

Theories about gender identity also generate claims about social roles and preferences that shift within this new environment. By understanding the process of identity construction, we may also understand the degree to which such roles and preferences are conveyed online. Central to this particular study is the notion that social roles and preferences are formed in large part before adulthood—specifically, during adolescence (Erikson, 1950; Lesko 2001). Much of media and technology usage is tangled up with adolescents' notions of the "real world." This age group is often acculturated into a patriarchal system of meaning-making, which is in large part an effect of the dominant mediated discourses surrounding them in conversations with parents and teachers (Finders, 1996), fashion magazines (Currie, 1999; Duke 2000; Duke & Kreshel, 1998; Durham, 1996; McRobbie, 1982, 1997; Milkie, 1994), romance novels (Christian-Smith, 1993; Rogers-Cherland, 1994), and the Internet and other computer-mediated technologies (Currie, 1999; Durham, 2001; Lewis and Finders, 2002; Sefton-Green, 1998; Stern, 1999). In romance fiction, women readers are provided "vicarious emotional nurturance" through identification with a fictional heroine whose identity as a woman is confirmed primarily by the sexual and romantic attentions of an idealized male hero (Radway, 1984, pp. 112–113). Adolescent romance novels offer cultural stereotypes of what constitutes young womanhood and constructions of femininity shaped by configurations of power and control (Christian-Smith, 1993; Rogers-Cherland, 1994).

While some adolescents resist these mediated discourses, the discourse of resistance is not the one that is most often articulated in schools or the media (Davies, 1993; Mazzarella, 1999). In other words, there are few discursive alternatives available that demonstrate differentially gendered positions in the worlds of adolescents. For example, boys are often steered toward careers and "logical" sciences while girls are steered toward consumerism and romance. This may perpetuate a notion among adolescent girls that they are constructed and validated through the (heterosexual) eyes of the opposite sex (Willinsky & Hunniford, 1993) and often drowns a resistant discourse. The result may be a culture of adolescence that is often confusing for girls who may seek alternative discourses from which to construct alternative identities or who wish to construct a comfortable identity while still attempting to fit in with peers and attain a media-perfect version of reality (Pecora & Mazzarella, 1999; Orenstein, 1994; Pipher, 1995). Some research demonstrates this opportunity to construct alternative identities—or even to become comfortable with the identities they have currently appropriated—through online communication (Sefton-Green, 1998; Turkle, 1995). Mediated communication provides a vast landscape upon which we can better understand adolescents' identity negotiations, but there is a particular need for study on the cultural narratives that manifest themselves through interactive media.

Identity and Gender: Exploring Social Roles as Construct and Performance

Gender, like identity, is definable on many parameters. Going back more than a half century, Simone de Beauvoir, in speaking of the identity of woman, focused on the enculturation process involved in "becoming" woman. She notes that women were not born but rather, they constructed themselves to become the gender of woman, or at least "woman" as the role was understood through the patriarchal cultural discourses of the time (de Beauvoir, 1949). In a more detailed explanation of what this means, Butler (1990) explains women are not born enacting a gender that is inherently feminine, but they learn to act like women by performing a culturally learned version of womanhood in a stylized repetition of acts from girlhood to adulthood.

Subjects construct and affirm identity through far more than the underlying meaning-making process about gender. In fact, it is thought by some to be impossible to discuss identity in only these terms. Identity should be considered through the lenses of gender, age, race, class, geographical location—really, any *particular* situation that in any way marks a person in a *particular* way—in order to be best understood in all its shifting intricacies (Abu-Lughod, 1990). All actions and texts are constituted around a particular discourse—or even multiple discourses (Foucault, 1974; Longhurst, 1989)—and "readers make sense of it in relation to the discourses (of age, race, gender, class, region, and so on) through which their consciousness makes sense of social reality and through which they are constituted as subjectivities" (Longhurst, 1989, p. 4). Therefore, while gender is paramount to this study of IM, these other factors are as important to this project and any study of identity.

The Internet as a Site for Identity Play

Since the inception of the Internet and the explosion of computer-mediated communication, scholars from various fields have sought to explore the sociological and cultural underpinnings of online identity formation. Much of the work has taken a cultural studies approach to investigate communication and culture within virtual communities and digital culture. The bulk of this work offers the theory that online conversations lead to the formation of distinct online identities that might be the same or entirely different from a person's "real-world" identity, which is based in part on psychological theory that assumes a "real" or core identity exists (Bromberg, 1996; Reid, 1991; Rheingold, 1993; Turkle, 1995). However, this work is at odds with the poststructuralist notion that there is no "real" core identity, but rather, many identities constantly shifting, depending on the cultural discourse being appropriated or negotiated at a given time; it should be noted that while psychologists—and much of the population outside academe—agrees with the notion of a core identity, this chapter takes up the poststructuralist view that identity is not fixed. More useful in this theoretical framework is Turkle's concept of "windowing," a widely cited metaphor designed to describe a person who "distributes" him or herself into multiple online conversations and acts differently among the conversations, taking on different roles all at the same time (Turkle, 1995). This is a common practice among IM users (Fabos & Lewis, 2000) and can seem very empowering. It posits that while online, limitations of real-world bodies may be overcome as well as its participants "by-pass the boundaries delineated by cultural constructs of beauty, ugliness, and fashion" (Reid, 1991, p. 42) and overcome the boundaries of gender, race, class, and age (Haraway, 1991; Reid, 1991).

In order to transcend, however, participants in computer-mediated communication first tend to explore *alternate* identities. While this means taking on an entirely different or a "masked" role within a conversation, research has demonstrated users just as often seek to maintain an identity consistent with their real-world identity (Irain Bowker, 2001; Lewis & Finders, 2002). Most often in IM or instant-relay chat, users adhere to a consistent presentation of self to acquaintances although they experiment with different tones, voices, and subject content among the different persons with whom they communicate (Cherny & Weise, 1996; Fabos & Lewis, 2000; Reid, 1991). Online identity construction and manipulation appears to be an enormous consideration in this communication process among adolescents. Negotiation is often located directly through the discourse of online communication—specifically through language use, social networking, and power negotiation among peers as well as a general surveillance of the social online landscape (Fabos & Lewis, 2000; Sefton-Green, 1998). However, scholars have found that rather than actually pretending to be someone else, the only place where some adolescents may feel comfortable expressing what they feel to be their "true" identity is in fact online (Tobin, 1998, p. 123), and they acknowledge a pureness (Clark, 1998) in their online relationships that is not present in their real-world ones.

While IM is very much a conversational medium, the idea that there is an audience for the text becomes paramount when trying to understand how identity is enacted when adolescents use it to communicate with one another. In theorizing the relationship between an audience and the text it reads, Hall (1980) has explained how encoding and decoding work: the text is encoded by the producer, and decoded by the reader, and there may be major differences between two different readings of the same code. However, by using recognized codes and conventions and relying upon audience expectations, the producers can position the audience and thus create a certain amount of agreement on what the code means. This is known as a preferred reading. In IM, this notion is challenged to a certain extent: The conversants sometimes appear to be talking in what indeed is a secret, teenage code rife with abbreviations and slang, but in the typing, each knows its audience will understand exactly what it is talking about. Because of this, the producer of the text no longer holds the privileged position of controlling the message; the power is divided equally between producer and audience because they are one and the same. The extremely high level of interactivity differentiates it from the mediated communications to which Hall referred, but IM also is quite different from other interpersonal and mass communication methods and technologies and is entitled to study in its own right. Its real-time nature differentiates it from email, which can be edited and reformatted endlessly in the same way that an old-fashioned letter or telegram might be. Its privileging of written text (text that can quickly and easily be copied and sent to different recipients) over oral conversation differentiates it from the telephone. IM conversations can be viewed as both a real-time exchange of information and ideas among a number of people and also saved and printed out for future viewing. They represent a certain blurring of private conversation and public space and, because of this, represent a particularly interesting landscape upon which real-world identities may be socially constructed in a virtual context.

In summary, this chapter attempts to understand how this intriguing, unique new medium might provide a site in which we can further understand identity construction among adolescent girls—the population that appears to rely most heavily upon it for daily communication (Lenhart et al., 2001). In doing so, I propose to answer two questions: What are the cultural narratives and representations used by girls in Instant Messaging? How does this online practice contribute to the construction of gender identity in adolescent girls?

Entering a Subcultural Space

Investigating the use of IM is complicated because the act of IM is usually solitary and private, relegated primarily to adolescents' homes and in the non-school hours. Often, they are online for hours carrying on these conversations, and printed, the conversations sometimes generate more than twenty pages per evening. Instant messaging, which is practiced most often in complete privacy, creates a subcultural space in which the identity process is enacted among adolescent girls. In an attempt to gain entrée to this subcultural space and better understand their thought process and creation of meaning through virtual conversation, I used a combination of in-depth and ongoing interviews with a dozen girls from different races and backgrounds and a narrative analysis of the conversations that they have with different people.[1] All girls, however, were avid users of IM who spent between ten and thirty hours a week on their home computers. In addition to having girls email me IM conversations (between 10 and 30 per girl over a period of several months) and doing in-depth interviews, I conducted ongoing IM conversations with a number of the girls to clarify and contextualize some of the conversations they sent me.

Furthermore, my methodology draws from the philosophy of feminist ethnography, a genre that calls for a combination of methods and acknowledgment of power structures within culture in order to create a "feminist ethnography landscape" (Bell, 1993, pp. 29–30). Feminist ethnography also takes into consideration the hegemonic nature of anthropology as a discipline rooted in the idea that a "researcher" studies and reports on (an often exotic) "subject," and attempts to dismantle this relationship by placing the studied subject on a more equal footing as the researcher (Behar & Gordon, 1995; Visweswaran, 1994). Most importantly, I will focus on the *cultural narratives* that manifest themselves both through the girls' conversations and their interviews. These narratives are integral to identity construction (Davies, 1993; Foucault, 1974; Gee, 1999). Through narratives, persons give meaning to their lives and construct themselves; narratives are crucial to shaping personal and social identity and crucial to understanding and constructing the identities of others, and this often plays out in very gendered ways (McLaren, 1993). Walker (2001) says all narratives are constructed discursively through the common "gender scripts" so that identities and processes of identification occur within the social networks and power relations that are most familiar in society (Walker, 2001). Munro (1998) claims studying narratives might "highlight gendered constructions of power, resistance and agency" (p. 7)—particularly shedding light upon the social relations that create and maintain gendered norms and power structures. Moreover, narratives have proven effective in gender communication research in large part because women's (and men's, for that matter) stories are often filled with latent meanings (Clair, 1993; Taylor & Conrad, 1992). Narratives are a crucial part of the identity process among children and adolescents. As Bronwyn Davies (1993, p. 17) writes in *Shards of Glass: Children Reading & Writing beyond Gendered Discourse*: "Each child must locate and take up as their own, narratives of themselves that knit together the details of their existence." She goes on to point out that:

> At the same time they must learn to be coherent members of others' narratives. Through stories we each constitute ourselves and each other as beings with specificity. (Davies, 1993, p. 17)

Upon reviewing the girls' IM conversations and interview data, I was able to discern a number of themes from the cultural narratives that occurred in each girl's conversations and interviews. I searched for themes, language, and signals that indicated what Gee calls "situated identity," or ways of "performing and recognizing characteristic identities and activities; and ways of coordinating and getting coordinated by other people, things, tools, technologies, symbol systems, places, and times" (Gee, 1999, pp. 38–40). Gee suggests asking a number of questions to help identify narratives and themes such as: "What cultural models are relevant here? What must I assume people feel, value, and believe consciously or not, in order to talk (write), act, and/or interact this?" "Are there differences here between the cultural models that are affecting espoused beliefs and those that are affecting actions and practices?" (Gee, 1999, pp. 78–79).

Although the girls studied all said they occasionally talk to friends on the telephone, their out-of-school communication primarily was through IM. Melissa, fifteen, who has been using IM for three years and still uses it an average of an hour per evening, said IM was a much more efficient way of reaching her friends— although much of this had to do with their phone lines being busy because of their constant Internet usage. Others said they prefer IM to the phone because, in the words of Beth, "the phone is boring." (This finding is consistent with the notion set forth by Fabos and Lewis (2000, p. 6) that phone conversations are riddled with "awkward silences.") Viewing IM as a preferred mode of communication among large sectors of adolescent girls, this analysis explores five themes that emerged from the cultural narratives in the girls' conversations and interviews. 1) IM is a space that is free from adult supervision and interference. 2) It is a means of constructing social status among peers. 3) IM is a technology that alternately allows adolescents to experiment with sexuality and gives them a more "normalized" space in which they can communicate with the opposite sex. 4) IM functions as a diary or journal in many ways and often seems to be the prime space for working through identity issues, such as religious conviction and sexuality. 5) Despite all the empowerment that might be found through online communication, the discourses in IM are largely patriarchal, and often the medium is yet another avenue for exclusion.

IM Is a Free, Unsupervised Space

Girls felt that their IM conversations took place in a "safe" or "free" space in which they could experiment with using different conversational norms than they do in real life, such as using profanity or asking more straightforward questions of their conversation partners. The girls manipulated the language they used in communication with others by altering tone, writers' voice, word choice, and subject matter of their communications more than what they admit in interviews they are likely to do in real life. This is evident in their online conversations when they use more punctuated, exclamatory language as well as profanity. Most girls in the study admit they would not use profanity in front of their parents or any adult, but as is apparent in the conversations, they all use it on IM. The ease with which girls may use profanity in the absence of parents and other authority figures is pertinent in a discussion of relevant cultural models within IM. In defining cultural models, it is important to answer the question of what feelings, values, conscious and unconscious beliefs are apparent within the conversations; in these particular conversations, the ease with which girls seem to transgress the hegemonic cultural models of femininity is worth examining. Girls in past studies adhered to a cultural expectation for niceness not just among adults but among other peers (Brown and Gilligan, 1992; Finders, 1996; Orenstein, 1994; Pipher, 1995). Because of unstated societal expectation that girls fit into unattainable models of perfection—from body size to politeness and even deference to male peers—the girls in these studies often took extreme measures to suppress negative or aggressive feelings and outward manifestations of these feelings. This longing to be the "perfect girl" appears to be changing in more recent literature, which paints modern girls as catty and aggressive with one another (Lamb, 2002; Simmons, 2002; Talbott, 2001). The strain of profanity and aggression found in these IM conversations not only flouts traditional standards of femininity, but it might also seem to exemplify a new way in which girls have embraced meanness with one another.

IM may also provide a space in which adolescents can experiment with language and abbreviated forms not condoned by the average English teacher. This includes the long-standing Internet chat abbreviations and phrases such as LOL for "laughing out loud" (which appears to be used as almost a stock response to a lot of phrases that are not necessarily all that funny), but it also includes abbreviation of standard phrases. For example, the phrase "not much" is often abbreviated as "nm," and "see you later" is generally written as "c u later." Conversely, one of the girls admitted in her interview that if a person does not adhere to these language norms in their IMing, they might be thought of, or even labeled as an outlier among her peers. Each girl admitted (and in communication on IM with me, demonstrated) that she wrote differently when communicating with adults by abandoning the abbreviations and slang terms and attempting to write in complete sentences.

IM May Confer Elevated Social Status

The girls used IM conversations to elevate their social standing in many interesting ways, a finding that is consistent with Fabos and Lewis' study (2000). Through the manipulation of tone and selective disclosure about themselves, they attempted to control conversation and present themselves in a positive light. For example, they disclosed private information and bragged about how many conversations they could hold at a given time in order to demonstrate clout and acquire social power. The girls studied mentioned that they would hold IM conversations with as many people as they possibly could. (Their "buddy lists"[2] contained as many as fifty other friends with whom they regularly chatted, and in some cases, many other names of people who were rarely if ever online). Numerous conversations, such as the following one between Beth and her friend Caroline, are entirely about how many people the girls are chatting with at a given moment.

Caroline says: *whos on for u?*
Beth says: *um. . lemme check*
Beth says: *anne, bethany jack,carna,u,colleen,amy,mandy ,an my cuzs friend*
Caroline says: *sh except i have more* [SH means "Same here]

Beth says: *like who??*
Caroline says: *jeff, Dylan, chris, carrie and alotmorw*
Caroline says: *and jack*
Beth says: *how many peeps r on??*
Caroline says: *10*
Beth says: *ic*

Girls also would take time before answering messages, even if they were not busy, to give the illusion of chatting with many people at a time. More subtle was the practice of using IM to make social arrangements—discussion of slumber parties, dances, and just hanging out in general seemed commonplace, probably just as commonplace as discussion of school and extracurricular activities. While IM provided a nice scheduling tool, the busier girls seemed to always be reminding others on IM just how busy they really were.

Change in Relational Dynamics, Experimentation with Sexuality

Another theme that emerged from this research was the notion that adolescent girls use IM to enhance relational dynamics and to experiment with notions of sexuality. Sexuality may be played out in various ways online, as is explored in Stern's (2001) study of girls' home pages, and in IM, one of these ways is through the use of log-in names and profiles that are attached to the log-in names. Although this study changes the girls' log-in names in the interest of privacy, most of the ones studied were in the form of a phrase that relegated some kind of identity to the person (whether true or false). For example, thirteen-year-old Leanne's log-in name for a week was "Leanne is SEXY!" and fourteen-year-old Madison used a sexually explicit line from a rap song as her log-in for a week. Often, sexuality is played out through the conversation itself, with topics ranging from breast size to comments about "hot guys." Much of the cultural discourse surrounding adolescents centers on the importance adolescents attach to sexiness and sexuality (Durham, 2001), and judging by the conversations among the girls—which often contained fights with boyfriends and ex-boyfriends and what some adults might consider sexual harassment from boys to the girls—IM does not appear to be a space that is free from this discourse. In conversations with various friends submitted by fifteen-year-old Aza, thirteen-year-old Liz, and fifteen-year-old Caitlyn, each in separate conversations jokingly referred to sexual lesbian acts in jest with boys with whom they conversed on IM. In some ways it is surprising (and perhaps a little bit heartening) that an alternative discourse to the dominant heterosexual norm is apparent in the conversations; after all, adolescents are so rarely offered alternative discourses within mainstream media and culture. However, these particular references to lesbians reflect a male fantasy of women with women that is sexualized and unrealistically depicted in pornographic films, and the discourse is ultimately negative and patriarchal. Here is an example of fifteen-year-old Caitlyn's boyfriend joking about the subject:

Luke says: *what did u do tonight*
Caitlyn says: *um i went to mcdonald's with anna at 7 till 8 then we went and got her moody brother and then she took us to randys*
Caitlyn says: *and think that aaron peters thinks me and marianne are lesbians*
Luke says: *lol*
Luke says: *sweet*
Luke says: *haha*
Luke says: *that would be cool*
Caitlyn says: *o no it wouldnt*
Luke says: *maybe it would be if u guys were bi*
Luke says: *lol*
Caitlyn says: *no no no*
Caitlyn says: *thats nasty*
Luke says: *not for me*
Caitlyn says: *boys are nasty*

However, their conversations also seemed to normalize the girls' friendships with boys. In participating in conversations with members of the opposite sex, Beth admitted she sometimes behaved differently than she might have in real life, either speaking more freely or acting more like a friend rather than "giggling." "Some things are just too embarrassing to say in real life," she said. Dominant notions of femininity still suggest that young girls should be more hesitant to broach delicate matters in as confrontational a manner as male peers, suggesting that girls are at best, "natural" mediators, and at worst, unlikely leaders. IM appears to provide an outlet in which girls may be as direct and confrontational as they like. In their interviews they said they felt more comfortable conversing in general via IM than in face-to-face or phone conversation. In the following conversation, Beth confronts a male friend about his relationship with one of her best friends—a confrontation she says she would not have had in person because she feared his face-to-face reaction:

Beth says: *so whats goi n on with u and bridget?*
Tyler says: *i dunno*
Tyler says: *do u no something that i shoud*
Beth says: *i dont think so besides that fact that u guys are like and never talk to each other*
Beth says: *i think personally u guys communacatied more when u guys werent a couple*
Tyler says: *watever*
Beth says: *whatever my butt*

In her interview, Melissa said IM provides her with a space in which she feels comfortable broaching sensitive topics and also allows her to chat with friends without the gaze of her peers to make her feel self-conscious. The following conversation with a male friend is the kind of conversation she said she would feel uncomfortable having in a public space:

Melissa says: *im kinda pissed @ u right now*
Bill says: *why is this*
Melissa says: *cuz lets see, u think the reason that i (transferred) to this school was because people said shit bout me*
Melissa says: *and its pissin me off*
Melissa says: *cuz u don't believe me that its not true*
Bill says: *when did i say that*
Bill says: *i just todl u what i heard*
Bill says: *i beleive u*
Melissa says: *and u continue to ask me about*
Melissa says: *and u asked dana bout, do obviously u didn't believe me*

This conversation demonstrates a sub-theme—the idea that IM questions the definitions of private and public space. While online conversation seems "private," its users have the ability to "cut and paste" it into another conversation with another person. Returning to Turkle's (1995) notion that the Internet is a disembodied medium that allows for play and experimentation, we should consider that much play and experimentation with identity might be on display for the world to see in much more apparent ways than a face-to-face conversation might demonstrate.

This happens in the next conversation when Melissa and Dana use multiple conversations taking place at the same time, specifically negotiating relational dynamics with a member of the opposite sex. Before the massive adoption of IM, adolescents often relied on telephone conversations to gossip about relationships, but through IM they can go directly to the source to check and spread the gossip. Cutting and pasting one conversation into another is a common means of accomplishing this. In the conversation on the following page, Dana cuts-and-pastes a conversation with a boy who is interested in Melissa and negotiates an indirect conversation between the two of them:

Dana says: *matt thinks your cute*
Melissa says: *smith???*
Dana says: *yea*
Dana says: *lol!!*
Dana says: *he likes you a lot*
Melissa says: *yeah he doesn't*
Dana says: *no for real*

Melissa says: *send me ur convo*
Melissa says: *pleaz*
Melissa says:
Dana says: *yea*
Dana says: *Here it is*
Matt smith says: *know who i kinda like. . . . .*
Dana says: *who is that???*
Matt smith says: *Melissa*
Dana says: *awwwwww*
Matt smith says: *but i doubt she would EVER go for me!*
Melissa says: *lol thats funny*
Dana says: *he really likes you*
Melissa says: *o my*
Melissa says: *that's funny*
Dana says:
Matt smith says: *tell her, im serious*
Melissa says: *lol, are u sending our messages back and forth??? lol lol lol lol*
Dana says: *not yours cause there all disses*
Melissa says: *did u send that to him??*
Melissa says: *lol*
Dana says: *hahah yea*

IM as a Diary

There are times when IM conversations may represent a private space, specifically when they are used as a diary. Each of the girls said she often saved and printed "important" or especially "funny" conversations in order to look back and review those conversations in the future: "I just sometimes like to remember what we talked about," Beth said, "but sometimes I want to keep it because it's funny."

Girls throughout history have used personal diaries as a means of articulating thoughts and feelings while negotiating identity (Brumberg, 1997), and IM conversations function in the same way as a diary—although they are articulated through interactivity with others rather than through introspection.

"I like to look back and remember what I was talking about with people and what I was saying," Melissa said. She said it's less to make sure everything is "on the record," but instead to better understand what was going on in her social life at a certain point in time. Aza agreed, saying she liked to look back on the funnier conversations, though she admitted she had just saved a few that represented the break-up fight of her and her most recent boyfriend. In this regard, communication through IM may be emerging as a primary source for how some adolescents conceive of themselves and take on an identity.

IM as Avenue of Exclusion
and Site of Dominant Discourse

Girls already mentioned that not adhering to the language-writing slang norms of IM might be grounds for exclusion. However, in the course of the study, one of the girls was a victim of being "blocked" by all of her classmates, who also happened to be doing the same thing at school by excluding her at recess, not inviting her to birthday parties, and making snide comments as they passed her in the hallway. IM allows its users to trigger a "block" mechanism that does not allow people to send messages to them; instead, a message is returned that tells them they have been blocked by the other or it masks that the user is even online. While this particular technological feature is indeed helpful in keeping away unwanted advances from IMing strangers,[3] it wound up being hurtful to the girl who realized what had happened. Although she said she felt "angry" and "like a loser" because of what was going on, the girl also said she felt thankful people were not coming up with new screen names to send her anonymous and taunting messages (a common practice in her clique, she said, and a practice mentioned by three other girls in their interviews).

However, it should be noted again that while it is a "free" technology, IM is only free to those who already have computers and working Internet connections. This excludes an enormous portion of the United States and world populations and can impact the entire social worlds of adolescents whose families cannot afford the technology but attend schools where using IM is an extremely important part of socialization. As Melissa said in her interview, students who don't use IM are not purposefully excluded while at school, but it is sometimes difficult for them to "catch up" on the social scene because so many important conversations (particularly those spreading gossip) take place on IM after school and in the evening. It stands to reason that adolescents with no financial means for IM at home would be constantly excluded from conversations that clearly go on long after the school bell rings at the end of the day.

Moreover, although the girls in the study often expressed themselves in ways that the girls in Brown & Gilligan's (1992), Orenstein's (1994), and Pipher's (1995) studies would have shunned (such as using vulgarity and sexual references and generally not appearing to be "nice" girls), these expressions were often mired in the dominant cultural discourses that cater to male desire and power. In her log-in name, twelve-and-a-half-year-old Leanne used "I really, really, really need a boyfriend!!!" for several weeks. Dominant cultural discourses relating to male desire are often played out in the discourses of IM conversations, such as this one between thirteen-year-old Liz and an ex-boyfriend that quickly dissolves from a chat about whether they are attending a classmate's bar mitzvah to a discussion of his online sexual exploits:

> Boy says: *i just had phone sex, . . .it was wierd*
> Liz says: *y do u tell me all this stuff?*
> Boy says: *i tell, ALL of my friends thats stuff, at least almost all*
> Liz says: *o*
> Liz says: *well*
> Liz says: *1 q: wit WHO?*
> Boy says: *oh, two girls that lives in (city), but i have a crush on one of them,*
> Liz says: *with TWO?*
> Boy says: *yeah:-)*
> Liz says: *omg . do they kno u r such a player?*
> Boy says: *i'm not. i don't try to get girls to sleep with me. ...wtf*
> Liz says: *wtf?*
> Boy says: *what the fuck*
> Liz says: *i don't kno wut to say*
> Boy says: *its not like you don't have options*
> Liz says: *wut should i say?*
> Liz says: *no 1 else i kno has phone sex wit a) multiple girls, b) ANYBODY!!*

Even though Liz clearly does not invite the discussion and seems disturbed by it (a fact that she confirms in a later email with a girlfriend), the boy provides lots of detail and seems intent on upsetting her. In an interview about the incident, she said she thought he might have been lying to hurt her (as they had only broken up a few weeks earlier), but she was not sure.

In addition to conversations about sexuality, many conversations among nearly all of the girls were concerned with looks and in particular, body weight. Fourteen-year-old Jenna used a screen name for several weeks that said (something to the effect of) "My butt is too fat." Conversations abound among girlfriends about weight and eating and did not change among different races and classes. This conversational preoccupation with appearance is consistent with much of the research on adolescent girls and their internalization of media discourses about attaining the "perfect" body (Durham, 1996, 1999; McRobbie, 1982).

And finally, it seemed that while girls were quite comfortable in transgressing gender norms by swearing and discussing sexuality, a large number of the conversations demonstrated the girls acting in the role of caretaker. Although this was rarely present in their conversations with girls, it was often apparent in conversations with boys—from Caitlyn talking her boyfriend out of driving drunk, to Jenna commiserating about weight issues with a male peer whom she had to comfort about feeling too fat, to Madison comforting a male friend over a girl who has stopped calling him. (The boys in these conversations never offered comfort to the girls, even when they complained about particular issues, such as Jenna with her weight.)

High-tech Note-Passing
or a New Paradigm for Identity Negotiation?

While some might argue that IM is really no different from the age- old act of illicit note-passing during class in terms of both the content of messages and the spirit in which they are quickly passed from one person to another, the technology demonstrates an immediacy and physical distance that cannot be overlooked in terms of its effects on identity and, in particular, gender identity. Physical bodies may be overlooked and forgotten on IM—even if not in actual representation of a person, then through the conversational tones and actions of that person.

A new generation of girls is constructing identity in large part without spoken words and personal contact. How exactly is this happening? In IM conversation after conversation, girls disclose different and often very contradictory bits of information about themselves; they appear to "try on" different tones with different persons. For example, fifteen-year-old Nikki IMs with a church camp friend and discusses the temptations of making out with boys and the "sinfulness" of wearing provocative clothing and lots of make-up (something she says she plans to give up immediately), but in a conversation the following day with a popular boy at school, she flirts, accepts compliments that she's a "hottie" and hints that she will indeed make out with him when they meet up over the weekend. And in the next conversation, later that evening with a male friend, she describes herself as shy and studious—very serious about school and church. While Nikki is very much a typical adolescent girl in that she must negotiate identity on so many fronts (sexuality, school, religion, appearance), it is fascinating and new that she can appropriate so many different identities in such a quick-and-easy way. However, this is all done without the benefit of face-to-face contact—and even without the blushing that might have occurred had she written and passed these notes along in class. Nikki, who does seem to be very shy and reserved in person, is able to experiment and shift her identity without self-consciousness about doing so, and she is able to do so at lightning speed.

While this in some ways may be very empowering for a group that historically has grappled with such issues as self-esteem related to body image, it is also raises questions about why/how dominant cultural discourses are carried into the online realm and the ramifications of this. For example, in Nikki's conversations, boys constantly told her she was "sexy" and "hot," and, while she sometimes accepted the compliments, she often played into the demure role that has been more expected of "nice" young women through history—an action that she admitted in her interview was generally reserved for boys because she was less comfortable being "forward" with them. In other conversations, girls of all races fret about their bodies (generally being too fat) and discuss how they plan to run and work out to make their bodies better, often in the place of sheer athleticism. Notions of aberrant bodies and male-dominated heterosexual discourses are prevalent throughout the IM conversations, a fact that might be surprising to those who believe the lack of face-to-face contact and physical body might grant equal footing to both sexes. The girls (and boys) appear to be very much a part of the cultural and mediated discourses that surround them constantly, even in this "free" realm of cyberspace. This falls in line with Butler's (1990) notion that gender is performed within the constraints of cultural discourses that people are unable to resist: within the IM conversations, girls and boys play into dominant cultural stereotypes and roles because even online, they know no other way to behave.

Furthermore, the very difficulty of recruiting a racially and economically diverse group of participants in the fieldwork signals a clear gap between those who have and use technology to enhance their lives and those who do not. Culturally, this gap—or "digital divide"—brings up issues of exclusion; the digital divide potentially silences not only individual voices in individual social milieus (like middle schools), but also entire groups within the public sphere.

Most parents and educators are concerned about IM in terms of whether it is a danger to their girls, and this question cannot possibly be addressed when we take into consideration all of the contradictory findings of this study: While girls feel empowered to break free from their attempt to appear perfect and nice by using foul language, saying what they wish, and experimenting with sexual tones in conversation, these actions often are steeped in the discourses that cast them as sexual beings who are weaker than men and ever in the male gaze. While the free space allows adolescents to experiment with language and develop newer means of communication, most of them stick to one preferred mode of communication and exclude those who do not write in the same way. And while IM permits girls to communicate with each other and with male peers without bodies and without the gaze of society, they often use the communication technology as another way to "block" or exclude peers. However, the main point of this study is to examine how identities are enacted in girls' communications with others. In this regard, it is clear that IM provides a ripe landscape for a girl to shift from identity to identity (for example, student to sexpot), and from moment to moment (as described by Hall, 1996), particularly when she carries on several different conversations at once. This is an opportunity for a girl to better understand who she is and *play* with who she wants to be in the future—an opportunity not afforded to past generations (and to lower social classes now).

Hugh Miller notes that in personal Web pages "information about the self is explicitly stated and can be managed by the person making the communication" (Miller, 1995). It seems that much the same phenomenon is happening with adolescents and IM, and in this way, despite the fact that their narratives tend to fit into dominant cultural discourses that seem harmful in terms of gender identity negotiation, it provides a means of agency. This supports the notion that some people feel more comfortable with themselves and articulating their identities on the Internet than they do in real life (Tobin, 1998; Turkle, 1995), and while some of the girls in this study often seem to willingly play into the discourses surrounding them in terms of sex and body image, they also are very willing to articulate their identities as strong, intelligent, and aware young women.

What this means for the adolescent girls as they continue to negotiate their identities both online and offline remains to be seen, but the prevalent use of this communication may change not only the ways in which girls communicate but the ways in which they conceive of themselves as they move into adulthood. If IM continues to burgeon as a preferred means of communication, the next generation (or at least certain classes within that generation) by and large will be a group that constructs and negotiates much of its identity—and in turn, gender roles and relations—primarily in the online realm. While this might signal the dawn of a brave, new world where inhabitants lock themselves in separate rooms in order to communicate with one another and construct identity as such, the findings of this study show that it is much more likely that, because we will all still be functioning within the confines of culture and prevalent discourses, things will remain very much the same.

Notes

1. The girls studied were primarily Caucasian, though two were African American, one was Korean American, and another was a Korean girl who had spent two years in the United States but had since returned to Korea. Their geographical backgrounds were somewhat diverse, with the bulk living in the Midwest (one in a college city; six in two different rural communities; two from the suburbs of Washington, D.C.; one from Baton Rouge, Louisiana; one from Chicago, Illinois; one from Los Angeles, California; and one from a suburb of New York in New Jersey). Their class backgrounds ranged from lower-middle to upper class with most falling in the middle. Each of the girls was very different from one another—some loved school; some hated it; some were popular; some were bullied by classmates; some were in single-parent homes and others had never seen their parents fight; some were very much into the idea of boyfriends while others seemed indifferent. All screen names were changed in order to protect the privacy of all involved.

2. In order to communicate with someone via IM, the user must have the other person's log-in name. Once they have this, they can add the name to a list of names of all of their friends, and when any of them are online, it appears in a little window on the screen and often is signaled with a computerized beep or the sound of a door creaking open.

3. When asked, most of the girls admitted they had been approached by strangers who had seen or guessed their IM login and sent them messages. While most said they ignored and "blocked" these messages, a few said that they would "play with" the person messaging them—sometimes flirting, sometimes by saying outrageous things. They then generally blocked the person if that person were a stranger.

References

Abu-Lughod, L. (1990). Can there be a feminist ethnography? *Women and Performance: A Journal of Feminist Theory, 5*, 7–27.

Althusser, L. (1968, trans. 1977). *Reading capital*. (B. Brewster, Trans.). London: Verso/The Gresham Press.

Behar, R., & Gordon, D. (1995). *Women writing culture*. Berkeley: University of California Press.

Bell, D. (1993). Yes, Virginia, there is a feminist ethnography. In D. Bell, P. Caplan, & W. Karim (Eds.), *Gendered fields: Women, men and ethnography* (pp. 28–43). London: Routledge.

Bromberg, H. (1996). Are MUDs communities? Identity, belonging and consciousness in virtual worlds. In Rob Shields (Ed.), *Cultures of Internet: Virtual spaces, real histories, living bodies* (pp.143–152). London: Sage.

Brown, L., & Gilligan, C. (1992). *Meeting at the crossroads: Women's psychology and girls' development*. New York: Ballantine Books.

Brumberg, J. J. (1997). *The body project: An intimate history of American girls*. New York: Random House.

Butler, J. (1990). *Gender trouble: Feminism and the subversion of identity*. New York: Routledge.

Chandler, D. (1998). Personal homepages and the construction of identities on the Web. Retrieved March 1, 2004, from http://www.aber.ac.uk/media/Documents/short/webident.html (Paper presented at Aberystwyth Post-International Group Conference on Linking Theory and Practice: Issues in the Politics of Identity, 9–11 September 1998, University of Wales, Aberystwyth).

Cherny, L., & Weise, E. R. (Eds.) (1996). *Wired women: Gender and new realities in cyberspace*. Seattle: Seal Press.

Christian-Smith, L. K. (1993). *Texts of desire: Essays on fiction, femininity and schooling*. NewYork: Routledge/Falmer.

Clair, R. P. (1993). The use of framing devices to sequester organizational narratives: Hegemony and harassment. *Communication Monographs, 60*, 113–135.

Clark, L. S. (1998). Dating on the Net: Teens and the rise of 'pure' relationships. In S. Jones (Ed.), *Cybersociety 2.0* (pp. 159–183). Thousand Oaks, CA: Sage

Currie, D. (1999). *Girl talk: Adolescent magazines and their readers*. Toronto: University of Toronto Press.

Davies, B. (1993). *Shards of glass: Children reading and writing beyond gendered identities*. Cresskill, NJ: Hampton Press.

de Beauvoir, S. (1949). *The second sex*. New York: Knopf.

Duke, L. (2000). Black in a blonde world: Race and girls' interpretations of the feminine ideal in teen magazines. *Journalism and Mass Communication Quarterly, 77*(2), 367–392.

Duke, L., & Kreshel, P. (1998). Negotiating femininity: Girls in early adolescence read teen magazines. *Journal of Communication Inquiry, 22*, 48–71.

Durham, M. G. (1996). The taming of the shrew: Women's magazines and the regulation of desire. *Journal of Communication Inquiry, 20*, 18–31.

Durham, M. G. (1999). Girls, media, and the negotiation of sexuality: A study of race, class, and gender in adolescent peer groups. *Journal of Mass Communication Quarterly, 76*(2), 193–216.

Durham, M. G. (2001). Adolescents, the Internet, and the politics of gender: A feminist case analysis. *Race, Gender & Class, 8*(3), 20–41.

Erikson, E. (1950). *Childhood and society*. New York: W. W. Norton.

Erikson, E. (1968). *Identity: Youth and crisis*. New York: W. W. Norton.

Evard, M. (1996). "So Please Stop, Thank You": Girls online. In L. Cherny, & E. R. Weise (Eds.), *Wired women: Gender and new realities in cyberspace* (pp. 188–204). Seattle: Seal Press.

Fabos, B., & Lewis, C. (2000). But will it work in the heartland? A response and illustration. *Journal of Adolescent & Adult Literacy*, *43*(5), 462–469.

Finders, M. (1996). *Just girls: Hidden literacies and life in junior high*. New York: Teachers College Press.

Foucault, M. (1974). *The history of sexuality*. New York: Vintage Books.

Gee, J. P. (1999). *An introduction to discourse analysis, theory and method*. New York: Routledge.

Gilligan, C. (1982). *In a different voice: Psychological theory and women's development*. Cambridge, MA: Harvard University Press.

Hall, S. (1980). Encoding and decoding in the TV Discourse. In S. Hall, D. Hobson, A. Lowe, & P. Willis (Eds.), *Culture, media, language* (pp. 197–208). London: Hutchinson.

Hall, S. (1996). Who needs identity? In S. Hall, & P. DuGay (Eds.), *Questions of Cultural Identity* (pp. 1–17). Thousand Oaks, CA: Sage.

Haraway, D. (1991). *Simians, cyborgs, and women: The reinvention of nature*. New York: Routledge.

Irain Bowker, N. (2001). Understanding online communities through multiple methodologies combined under a postmodern research endeavour forum [Electronic version]. *Qualitative Social Research*, *2*(1).

Kendall, L. (1996). MUDer? I hardly know HER!: Adventures of a feminist MUDder. In L. Cherny, & E. R. Weise (Eds.), *Wired women: Gender and new realities in cyberspace* (pp. 216–217). Seattle: Seal Press.

Lamb, S. (2002). *The secret lives of girls: What good girls really do—Sex play, aggression, and their guilt*. New York: Free Press.

Lenhart, A., Rainie, L., & Lewis, O. (2001). Teenage life online: The rise of the instant-message generation and the Internet's impact on friendships and family relationships. Washington, DC: Pew Internet and the American Life Project. Retrieved April 29, 2004, from http://www.pewinternet. org/reports/toc.asp?Report=36

Lesko, N. (2001). *Act your age! A cultural construction of adolescence*. New York: Routledge/Falmer.

Lewis, C., & Finders, M. (2002). Implied adolescents and implied teachers: A generation gap for new times. In D. E. Alvermann (Ed.), *New literacies and digital technologies: A focus on adolescent learners* (pp. 101–113). New York: Peter Lang.

Longhurst, D. (1989). *Gender, genre and narrative pleasure*. London: Unwin Hyman.

Mazzarella, S. R. (1999). The "Superbowl of all dates": Teenage girl magazines and the commodification of the perfect prom. In S. R. Mazzarella, & N. O. Pecora (Eds.), *Growing up girls: Popular culture and the construction of identity* (pp. 97–112). New York: Peter Lang.

Pecora, N., & Mazzarella, S. R. (1999). Introduction. In S. R. Mazzarella, & N. O. Pecora (Eds.), *Growing up girls: Popular culture and the construction of identity* (pp. 1–14). New York: Peter Lang.

McLaren, P. (1993) Border disputes: Multicultural narrative, identity formation, and critical pedagogy in postmodern America. In D. McLaughlin, & W. Tierney (Eds.), *Naming silenced lives—Personal narratives and the process of educational change* (pp. 201–235). New York: Routledge.

McRobbie, A. (1982). *Jackie*: An ideology of adolescent feminism. In B. Waites, & T. Martin, *Popular culture: Past and present* (pp. 263–283). London: Open University Press.

McRobbie, A. (1997) *Back to reality? Social experience and cultural studies*. Manchester: Manchester University Press.

Milkie, M. (1994). Social world approach to cultural studies: Mass media and gender in the adolescent peer group. *Journal of Contemporary Ethnography*, *23*(3), 354–380.

Milkie, M. (1999). Social comparisons, reflected appraisals, media: The impact of pervasive beauty images on black and white girl self concepts. *Social Psychology Quarterly*, *62*(2), 190–210.

Miller, H. (1995, June). The presentation of self in electronic life: Goffman on the Internet. Retrieved March 9, 2004, from http://ess.ntu.ac.uk/miller/cyberpsych/goffman.htm (Paper presented at Embodied Knowledge and Virtual Space conference, Goldsmiths' College, University of London).

Munro, P. (1998). *Subject to fiction: Women teachers' life history narratives and the cultural politics of resistance*. London: Open University Press.

Orenstein, P. (1994). *Schoolgirls: Young women, self-esteem, and the confidence gap*. New York: Doubleday.

Pipher, M. (1995). *Reviving Ophelia: Saving the selves of adolescent girls*. New York: Ballentine.

Radway, J. (1984). *Reading the romance: Women, patriarchy and popular literature*. Chapel Hill, NC: University of North Carolina Press.

Reid, E. (1991). Electropolis: Communication and community on Internet Relay Chat [Retrieved Online]. Unpublished honor's thesis, University of Melbourne, Australia. Retrieved on March 9, 2004, from http://eserver.org/cyber/reid.txt

Rheingold, H. (1993). *The virtual community: Homesteading on the electronic frontier*. Reading, MA: Addison-Wesley.

Rogers-Cherland, M. (1994). *Private practices: Girls reading fiction and constructing identity*. London: Taylor & Francis.

Sefton-Green, J. (1998). Introduction: Being young in the digital age. In J. Sefton-Green (Ed.), *Digital diversions: Youth culture in the age of multimedia* (pp. 1–20). London: University College London Press.

Simmons, R. (2002). *Odd girl out: The hidden culture of aggression in girls*. Orlando, FL: Harcourt Inc.

Stephenson, W. (1988). Play theory of mass communication broadly considered. In W. Stephenson (Ed.), *The play theory of mass communication* (pp. 190–206). New Brunswick, NJ: Transaction Books.

Stern, S. R. (1999). Adolescent girls' expression on web home pages: Spirited, somber and self-conscious sites. *Convergence, 5*(4), 22–41.

Stern, S. R. (2001). Sexual selves on the World Wide Web: Adolescent girls' homepages as sites for sexual self-expression. In J. Brown, J. Steele, & K. Walsh-Childers (Eds.), *Sexual teens/sexual media: Investigating media's influence on adolescent sexuality* (pp. 265–286). Hillsdale, NJ: Lawrence Erlbaum.

Talbot, M. (2002, February 24). Girls just want to be mean: Mean girls and the new movement to tame them. *New York Times Magazine*, 24–29, 40, 58, 64–65.

Taylor, B., & Conrad, C. (1992). Narratives of sexual harassment: Organizational dimensions. *Journal of Applied Communication Research, 17*, 401–418.

Tobin, J. (1998). An American otaku (or, a boy's virtual life on the Net). In J. Sefton-Green, (Ed.), *Digital diversions: Youth culture in the age of multimedia* (pp. 106–127). London: University college London Press.

Turkle, S, (1995). *Life on the screen: Identity in the age of the Internet*. New York: Simon & Schuster.

Visweswaran, K. (1994). *Fictions of feminist ethnography*. Minneapolis: University of Minnesota Press.

Walker, M. (2001, March). Engineering identities. *British Journal of Sociology of Education, 22*(75), 15–30.

Willinsky, J., & Hunniford, M. R. (1993). Reading the romance younger: The mirrors and fears of a preparatory literature. In L. Christian-Smith (Ed.), *Texts of desire: Essays on fiction, femininity and schooling* (pp. 87–105). London: Falmer.

11

The Constant Contact Generation

Exploring Teen Friendship Networks Online

Lynn Schofield Clark

Introduction

When sixteen-year-old Steph Kline[1] was asked to describe her uses of instant messaging technology,[2] she could barely contain her enthusiasm:

> Steph: *I can talk to up to 10 people at the same time. You hear a little blinker and it comes on. A new way of communicating!*
> Interviewer: *Is that easier for you than using the phone?*
> Steph: *Yeah, it is. And also it makes your typing a lot better and you need that in school because it is an essential skill that you need. It helps a lot. I can express myself a lot better when I write than when I talk. Plus it is just a lot of fun. You don't know what to really expect.*
> Steph's mother: *But they call each other, too!*
> Steph: *I do use the phone a lot too.*

In this conversation with her mother and an adult interviewer,[3] Steph spoke of the multiple simultaneous communications she had with her friends online. She also deftly wove in a nod to the "essential skill" of typing and of written communication fostered through her use of this technology, highlighting for her adult audience the educational benefits of her online practices.

I begin this chapter with this excerpt because it highlights several important ways in which new media have come to play a part in the everyday lives and the interactions of U.S. teen girls with their parents and their peers. In this and in other passages in their interviews, Steph's mother and Steph highlight the differences between communicating by cell phone and computer in the home. Whereas cell phone conversations are seen as a disruption to homework, computer-related communication is viewed more ambiguously. Steph's comment about improved typing skills implicitly recognizes the dual role of computer technology in the home, in that the computer is employed by teens for both educational and entertainment-oriented leisure pursuits. In Steph's comment, we see how one teen is able to rely on this duality, and the ambiguity of computer use that results, to subtly garner her mother's support for her use of an "educational" technology. Yet through instant messaging and email, pleasant interruptions to homework, that most educational of the uses of the technology, might come at any time from anyone. Seated at the computer, Steph was able to maintain the appearance of a solitary focus on her homework, thus keeping her parents happy, while simultaneously keeping up her communication with friends.

A second point illustrated in Steph's comment occurs as she mentions the fun of using the technology and of not knowing what to expect. This is a reference to the fact that differing people might be online at different times, thus changing the dynamics and possibilities of various peer-to-peer conversations. Steph enjoyed expressing herself through writing, thus making her online experience a context in which she could gain confidence in herself and in her relationships.

Not all teens share the positive experiences with the online environment that Steph related. Not all have the same level of access to the technology as Steph did, either (Facer and Furlong, 2001). Yet Steph's story expresses some themes that were common among the 44 teens and 12 "tweens" involved in the in-depth interviewing that formed the basis of this chapter, as all of those teens, and others in the U.S., as well, are shaped within the same cultural developments that have transformed modern societies in recent years. The chapter, therefore, is framed by Anthony Giddens' (1990) theory of "radicalized modernity" and James Slevin's (2000) application of Giddens' theories to the phenomenon of the Internet. Thus, while my discussion deals primarily with those teens who talked in the context of in-depth interviews about their uses of the new technologies, the chapter employs this data to explore the ways in which these practices grow out of and are intimately related to the demands of society and of life as these teens experience it within their particular peer-oriented contexts. In seeking to understand this process, the chapter presents what I term the phenomenon of *constant contact*, a principle that organizes the uses of new media among teen girls. I explore why peers seem to desire such unceasing contact with one another, the benefits and risks that occur as a result, and the implications of these practices for teen/adult relations and for society in general. Because teens use not only Instant Messaging, but also short messaging services (SMS), multimedia messages (MMS), email, and cell phones within this context of constant contact, all of these technologies will be considered.[4]

Placing Individual Use
in the Peer Group Context

According to a recent study by the Pew Internet & American Life Project, almost seventeen million young people between the ages of twelve through seventeen use the Internet (Lenhart, Rainie, & Lewis, 2001). Nearly half of the teens surveyed said that they believed that the Internet improved their relationships with their friends, and a third said that it helped them to make new friends. Three-quarters of online teens said that they believed the Internet helped them in school. Three-quarters also reported using instant messaging, with most using it at least several times a week. The same study also found that more than half of parents with online teens thought that it was essential for young people to learn how to use the Internet in order to be successful, and almost 90 percent believed that the Internet had helped their children in school. Clearly, Instant Messaging, email, and the resources available through the World Wide Web have found an important place in the lives of contemporary teens.

While these uses seem fairly commonsensical, the current uses of the new technologies can be understood in a broader historical framework. Indeed, it is possible to argue that focusing on these individualistic, instrumentally understood uses of the new technology can mask the deeper and less conscious motivations that have shaped new media use on an individual level, while also masking the economic forces that place the technologies within use to begin with.

Placing the social theories of Giddens within the context of the Internet, James Slevin (2000) has argued that the new technologies have become useful to people as they aim to stabilize the outcomes of the complex social transactions in which individuals and groups are now forced to engage in late modernity. From biotechnology to online governance to market analysis, technology is employed to minimize some of the risk inherent in decisions at the individual and social levels. Yet the new systems that have emerged unintentionally introduce new forms of uncertainty and risk. As Slevin notes, "there are no taken-for-granted solutions and the individual has to develop skills in order to gather together and process this information, weighing up the likely risks and certainties related to different outcomes. Modern everyday life is rife with such moments" (p. 16).

While Slevin speaks mainly of adult dilemmas, teens also are confronted with less specific paths for their own life courses and might be expected to utilize new technologies to similarly minimize risk. As part of their coming-of-age tasks, young people need to develop the skills necessary to envision various possible out-comes to their actions. Their choices are often shaped by occurrences over which they have no control, such as a parent's divorce, job changes or losses, changes in their family's financial picture, or even changes in their family's geographic location. Yet it is important for young people to believe that they are exerting some measure of power over their own environment in a desire to minimize risk. This theme of seeing oneself as powerful within particular contexts is manifested in two ways in this chapter: first, as teens employ various tactics to work within the limitations and expectations placed upon them by their parents, and second, as they attempt to con-trol various online interactions so as to both maintain relationships with and to ensure that they remain favorably evaluated by their peers.

Certainly, the process of consciously pursued information-gathering is an important aspect of the attempt to minimize uncertainty for teens, and a great deal of research has explored the ways that people of differing ages evaluate information-al materials online (Dresang, 1999; Hakken, 1999, 2003; Hirsch, 1999; Kuhlthau, 1991; Scanlon and Buckingham, 2003). Yet this chapter asserts that there are also less formal and often unconscious ways in which teens seek to exercise mastery over their own lives through the use of the technologies available to them. As will be demonstrated, these less-conscious practices might better be understood in relation to what sociologist Ann Swidler (1986) has referred to as "strategies of action" (pp. 273–286). Swidler argues that "strategies of action" are not the result of an individ-ual's conscious, rational choice-making. Rather, she argues, people commonly oper-ate out of an often taken-for-granted understanding of how things should be done. People acquire this sense of how things should be done mostly as a result of their relationships with other people. This, in turn, has a great deal to do with where they are located socioeconomically, geographically, and in terms of such categories as age, status, gender, racial/ethnic group, and so on. For teen girls, I argue that strategies of action emphasize peer-related processes of evaluation, such as the subtle nego-tiations concerning everything from appropriate attire to appropriate academic and career aspirations. Thus, uses of the new technologies need to be understood within this peer-dominated framework as the technologies facilitate not only increased opportunities to obtain abstracted information but also opportunities for constant contact with their peers, as well.

Previous research has explored the ways in which teens and adults view new technologies in similar and in different ways. In their study of cell phone uses by adults and teenagers in Norway, Ling and Yuri (2002) found that both parents and teens used cell phones for "microcoordination," but teens alone used it for what they termed "hypercoordination" (p. 139). Whereas microcoordination references the ability to use the cell phone to adjust meeting times and places at a moment's notice, hypercoordination refers to additional, expressive uses of the phone, such as in-group discussion of appropriate phones to use, peer norms of use, and how each relates to presentation of self. Teens were concerned with not just how the technology could facilitate constant contact with their peers, but they were also concerned with what such contact itself communicated. The teens interviewed talked about which phones were deemed acceptable by their peers and which were not. They discussed the appropriate norms concerning answering immediately, requiring the caller to leave a message, or blocking calls. Moreover, the researchers noted that whereas adults felt stressed by the additional demands incurred through mobile phone use, teens "thrive on access and interaction. To receive a message is a confirmation of one's membership in the group" (Ling & Yuri, 2002, p. 149).

Leonardi's (2003) research is also suggestive of differences in how parents and teens interact with the technology. In his study of first-generation Latino working-class immigrants and their perceptions of cell phones, computers, and the Internet, Leonardi found that many adults believed that the Internet and computers could provide much to family life that was positive. When these technologies were believed to distract users from interpersonal activities within the home, however, they were viewed much more negatively. Overall, Leonardi argued, his interviewees tended to view communication technologies as promoting an ethic of individuality that served to remove users from the social life of the present moment.

But such removal from the present moment is often exactly what teens are seeking when they employ new media technologies in the home. As Ling and Yuri (2002) noted in their study of cell phone use in the home, "the mobile telephone taps into (the need for teens to establish an identity apart from their families) in that it provides adolescents with their personal communication channel . . . (it) has the advantage of being outside the purview of authority figures" (p. 152). The cell phone, instant messaging, email, and text messaging all allow the teen to be located in the home while also remaining in contact with her peers.

It is important to note that for instant messaging, email, and cell phone use, communication with other persons, rather than consumption, is the primary motivation for participation. Of course, even this interpersonal communication participates within the broader interests of several interrelated communication industries (Kasesniemi & Rautianen, 2002). They also serve these interests, albeit in indirect ways that often go unrecognized by the teen users themselves (Clark, 2003). This chapter therefore aims to explore the connections between economic interests and the desire on the part of individual users to see themselves as the ultimate decision-makers regarding how the technology might be employed in their relationships with their parents and their peers.

Getting Beyond the "Official Response"

Much of the research into how Web use might be understood in relation to life offline has been focused on one of two phenomena. First, "virtual ethnography" research has explored the construction and use of "virtual communities," a body of research that has provided a great deal of insight into fan-based communities, subcultural and diasporic identity maintenance, and various practices of consumption (Baym, 2000; Jones, 1995, 1998, 2000; Turkle, 1995). Second, research has explored how individual information-seeking online might be related to the desire for engagement in civic life (Corrado & Firestone, 1996; Norris & Jones, 1998; Rice & Katz, 2004; Shah, Kwak, & Holbert, 2001). Whereas the former has been rooted in in-depth explorations of the online environment, the latter has been gleaned from survey data that has sought to identify interrelated variables.

This chapter is based on a third approach that takes as its starting point the offline context of teens' everyday lives and their interactions with one another both off and online. Through narratives gathered in in-depth conversations with teen girls and boys interviewed on their own and within their families and friendship circles, the chapter explores how the World Wide Web provides a contemporary context for the longstanding teen desire to be in constant contact with their peers and to exercise some control over their relationships with their parents. It therefore shares methodological ground with similar projects taking place in Canada (Shade, 2002), Denmark (Tufte, 2003), and in the U.K. (Buckingham, 2000; Livingstone, 2002; Livingstone & Bober, 2003; Livingstone and Bovill, 1999).

Part of the rationale for employing in-depth interviewing grows out of a desire to begin the study with attention to how young people themselves explain what is important to them about their new media use. As Facer, Sutherland, Furlong and Furlong (2001) have noted in their studies of young peoples' uses of new media, it is important to get beyond the "official response" to questions that teens know they are expected to provide to adult audiences. This requires the development of trust between the interviewer and interviewee, which is established in part by the interest an interviewer takes in the teen's whole life (in other words, not just in his or her educational achievement in the school setting).

In this study, 44 teens (young people aged eleven to eighteen) and 12 "tweens" (young people aged eight to eleven) were interviewed with their families, individually, and in some cases, among a group of their peers. They were asked to discuss their uses of various technologies inside as well as outside of school. New media use in relation to their friendship circles quickly emerged as the most salient of uses. Interviews were conducted by a team of interviewers between June 2001 and November 2003 and were analyzed in an ongoing group discussion among faculty members and graduate students who shared a research interest in teens, in conjunction with conversations among colleagues conducting similar research in other locations.

The writing of this chapter takes the form of experimental qualitative research, in that rather than excerpting certain relevant passages and abstracting them from their contexts, the chapter opts for a case-study approach. Thus, the chapter maintains the flow of a singular story through emphasis on a detailed examination of one case, highlighting throughout the ways in which that one story connects with those of other young people interviewed. We therefore continue the story of Steph Kline, introduced at this chapter's outset.

Case Study: Steph Kline

The home computer, teens, and parental supervision

Steph Kline was the most computer-savvy member of her middle-class family. Her mother, a Latina born in Guatemala, and her father, an Anglo-American, both had little experience with computers in their work worlds, and her younger brother had found little use for them. Steph, an honor student at school, was very involved in community and school theater, often taking the lead role in musicals. A chatty, articulate and attractive young woman, Steph juggled a demanding school curriculum, extracurricular activities, and a part-time job, prioritizing contact with her friends through the use of her cell phone and the family's Internet connection at home. The computer was therefore primarily a way for her to be in contact with her friends while she was located in her home.

Viewing the home computer as a means to multitask—to do both homework and to maintain communication with peers—was common among most of the teens interviewed. In fact, most took pride in their ability to do many things at once. Those teens who did not have a computer in their home spoke of similar multitasking, although this took place in libraries, schools, or at the homes of grandparents, other relatives, or friends. In these situations, with their inherent limitations of time and access, emailing and Web surfing were more attractive than instant messaging. Sometimes cell phones and short messaging services worked as alternate ways for teens to be in touch with peers when in homes without computers.

It is important to point out that these practices of constant contact with peers were directly related to contextual factors over which most teens (particularly the younger ones) had little control. Increased concerns among parents about safety, combined with the pressing parental needs of work, housekeeping, and in many cases the demands of single parenting, means that today's teens are often required to spend more time in their own homes than was the case in generations past. As Livingstone (2002) points out, we need to consider the context of childhood as not democratized but bureaucratized. It is a highly scheduled context. In Steph's case, as well as for many young people in the U.S., life is organized around schoolwork that is to be completed in one's own home. Given the choice, most young people would prefer to be with their friends in unstructured, unsupervised activities. Being in contact with one another via technological means to discuss homework as well as other matters, therefore, has evolved as a logical second choice for teens, given the contexts of their lives. It is a choice with its own set of challenges in relation to both the context of the family and one's friends. One of these challenges arises in the issue of parental supervision.

In the Kline home, the computer sat in the back of the formal living room, near the kitchen and family room. It was moved there from the basement, Steph's father explained, because Steph's parents "wanted to monitor Steph's Internet and chat use." "She was spending a lot of time on the computer and away from us," her mother added, "and we did not know what was going on and I heard a lot of things about teens getting into things on the computer." Steph's mother related her fear that her daughter could become involved in unsavory activities or relationships through the online environment. The solution, according to Steph's mother, was to take an active role in the lives of Steph and her younger brother, which included knowing their friends and being aware of their activities both online and off.

While Steph noted in an individual interview that she had become more focused on homework after the computer had been moved to the living room where she normally did her homework, she found that using the Internet to communicate with friends during homework sessions had several practical functions when compared with the use of the family phone line. For one thing, it afforded the possibilities for communication in a group rather than solely in a one-on-one context. This ability for group communication seemed to be especially appealing to girls. Yet Steph noted that group emails and group instant messaging sessions were not always the best ways to tackle homework assignments. When conversations about homework or even about social concerns became complicated, she preferred to use her cell phone. For everyday communication, however, an unstated advantage of instant messaging for Steph was not having to worry about expiring cell phone minutes.

The primary advantage of instant messaging, however, had to do with Steph's parents' desire to supervise her activities and her own desire to exercise control over how that supervision would take place. As she said, "I mean, on the phone they can hear you but when you are on the Internet all they can hear is you typing." While Steph said that her parents "always want to read all my conversations," she protested, "I'm not doing anything wrong but it is just that I want to talk to my friends." Like most teens, Steph valued her privacy, and Internet-based conversations afforded her opportunities to retain this privacy within a public space in her home.

For the most part, Steph considered the computer to be her own. At the time of the interview, the family was experiencing some tensions over the fact that they had only one phone line. This meant that when Steph was online, no one in the house could receive or make phone calls. Steph's parents had said that their friends and business associates complained to them that they often could not get through because the line was frequently busy. As a result, her parents were considering investing in a DSL line.

The parental desire to provide supervision over Steph's activities online, which had necessitated the move of the computer to a more public place in the home, had also resulted in Steph's increased use of the technology. The unintended consequence was the inconvenience of a frequently busy phone line and the prospect of an expensive DSL line. Through her use of new media technologies, Steph therefore had indirectly negotiated an increased ability to both control her parents' supervision of her activities and pursue the contact with friends she desired, all at her parents' expense.

Communication with peers

Like most teens we interviewed, Steph was not interested in going into chat rooms where "random people" could enter, as she described it. She preferred to create her own chat rooms where she could interact with her friends. Thus, the Internet has been a way for her to deepen relationships with friends from her offline life, and for her to manage those relationships in ways that indirectly worked to her advantage. Consistent with her self-assessment of her written communication skills, Steph believed that her online conversations were "deeper" because she could express herself better, in part because communicating online afforded her the opportunity to plan out what she wanted to say in advance. Instant messaging and email conversations on the Internet were viewed by Steph as a supplement to the contact she had with her friends in the school corridors and on the phone both during and after the school day.

Fourteen-year-old Annae Gardner was another teen girl who, like Steph, saw advantages to the written character of online communication and the ability it affords for controlling both one's own emotions and one's presentation to others.[5] A working-class Anglo teen, who tended to be an outlier in relation to her peers, Annae noted that communication online gave her a chance to control her temper. When something said online made her angry, she said that she went to her room and cursed and stomped around until she felt calmer. Then, she returned to the online communication and was able to avoid saying something she would later regret. Other teen girls, such as Montana Odell, a popular biracial sixteen-year-old from a relatively well-off background, noted that she sometimes received emails from people she did not know very well who might feel shy about talking with her in person.[6] It seemed that in the case of Montana as well as a few other teens, the Internet allowed girls to initiate conversations online that might blossom into friendships, and thus the written communication online provided a less socially risky encounter than might be possible when seeking out someone in school.

Given her strong expressive skills, I was particularly interested in whether or not Steph had had conversations online that she believed might not have taken place in the offline environment, similar to those mentioned by Montana Odell. In a follow-up interview, Steph said that the Internet afforded her with important opportunities that would not otherwise have been available to her and her friends, but it had also increased the possibilities for miscommunication:

> Interviewer: *Has the Internet ever created a problem that you then had to go work out in a face-to-face situation?*
> Steph: *Yeah.*
> Interviewer: *Really? Oh, do tell.* (Both laugh).
> Steph: *Well, I know a lot of times, I am kind of known as "the psychiatrist" because a lot of people come to me naturally for their problems. I love having people tell me things and then I listen and give my input. I just love doing that. I just love hearing people's problems and stuff and helping them. So, I notice that a lot of times I will just be doing different things and my friends will IM and a window will pop up and say "I need help" and they will start the conversation by saying something . . . sometimes it could be they have a problem with me or something. Like, they might say, "I really didn't like the way you handled this situation today." And, I think it is easier for them because they are not talking to me face-to-face and then afterwards, when I realize it is a bigger issue that shouldn't be handled just on the computer, then I will say "we need to talk tomorrow, face-to-face." So we will set up little counsel sessions and stuff so that we can really just talk face-to-face but the computer definitely opens up the initial conversation. Because I think people get scared in bringing up a topic of conversation and kind of just walking up to someone saying "oh, I've got a problem with this." I think you can hide behind the computer and say it with words. But then, yeah, I think it definitely comes down at times to having to go off the computer if it is a bigger issue.*

Steph, like other teen girls interviewed, viewed online communication as an extension of relationships that were an important part of her day-to-day life. Thus, sometimes the computer was helpful as something to "hide behind," as she said. As such, the computer lowers risk and makes possible talk about topics that might be uncomfortable subjects for discussion in a more immediate, in-person situation.

> Interviewer: *How do you know when something is right for the computer or not right?*
> Steph: *I think, mostly just because some conversation is for the computer like homework-related things but I think if it is more like issues that are going on with someone like . . . I mean it can go as far as "I'm thinking of hurting myself because something happened," or "my boyfriend dumped me today." I think issues that are really affecting people, those kind of problems, then we would want to really talk about it.*

According to Steph, such a lowering of risk might have a real value in cases such as when a friend alerts another friend to her pain. It might be easier to reveal to a friend online one's temptation to hurt oneself, Steph implied. After that revelation, an in-person conversation would be necessary.

Not all teens agreed with Steph, Annae, and others about seeing the written demands of online communication as a benefit, although more of the girls than the boys seem to see it in this way. Fifteen-year-old Seth Rodanski, for example, confessed that his incompetence in typing and in writing made in-person and phone communication far preferable to online communication, a sentiment echoed by numerous other boys, although many boys noted that they mostly preferred to compare notes about gaming conquests online.[7] Some boys, such as five fifteen- to seventeen-year-old boys in one friendship circle, expressed discomfort at the idea that they would have extended conversations with other males.[8] Consistent with this, the boys' appreciation of the "face-saving" element of online communication was also specific to gendered norms: rather than feeling they might be privy to the more personal and even troubling thoughts of others, as Steph had noted, Jacob and Marcus Tabscott liked the fact that if they made someone angry; they had fewer fears of repercussion. As Jacob said, "If you know they're in like Kansas somewhere," they cannot come and beat you up, or as his friend Caleb Baylor added, "you don't have to see them if they're laughing at you." While the girls did not share the boys' discomfort concerning same-sex communication, there was evidence among both boys and girls of the heterosexist norms online that permeate U.S. high school life, as well (see Clark, 1998).

There are also problems that can emerge as a direct result of the technology's written nature. As Thiel (2005) notes in this volume, girls use the cut-and-paste function to pass along to a third party purportedly confidential communication between two people. One teen interviewed in this project, thirteen-year-old Gabriella Richards, recalled a controversy among her peers when someone printed out such a confidential message and passed it around to others, destroying the note-writer's credibility.[9] Written communication has long carried inherent risks. The email message, like its note-passing predecessor in teen culture, signifies a private conversation yet holds the risk of revelation to others.

Many of the teen girls interviewed, such as sixteen- and seventeen-year-olds Miranda Murguia and Victoria Benitez, employed the shortcut of forwarding "chain" emails as a means of communicating with their peers.[10] These and other teens interviewed talked of sharing illicit jokes, passing along quizzes about love interests, or sending saccharine poems or greetings about friendship as a means of being in contact with their friends. While they seemed to appreciate the fact that they could avoid the pressure of composing their own messages, other teens, including both boys and girls, expressed frustration at the demands within many of these notes to "pass this along to fifty friends."

The issue of numbers is important when it comes to email, instant messaging, and the mobile phone. Steph, as well as the vast majority of other teens interviewed, tended to view email, instant messaging, and mobile phone technologies as a means by which to quantify, measure, and verify one's popularity with one's peers. Steph noted that she used three different instant messaging services at one time so as to keep several conversations going at once. While Steph proudly stated that she was able to talk to ten people at one time, thirteen-year-old Jeremy Friedman proudly noted that he has held conversations with twenty-five different people at the same time using instant messaging.[11] Other teens bragged of how many people they had listed in their online address book and in their cell phone's speed dial directory, with some noting numbers in the hundreds. Many knew the upward limits of their address book or directory and strove to reach those limits, necessitating the occasional elimination of contact information for some less-frequently contacted persons. Fifteen-year-old Seth Rodanski even told of introductions he had made between his cousin and his own friendship circle so that she could expand what he considered to be her paltry list of instant messaging buddies.

A finding that Ling and Yuri (2002) note in relation to Norwegian teens and their use of the mobile phone to quantify popularity holds true in this study as well: having large numbers of contacts in one's instant messaging and mobile phone directories is a symbol of popularity. Even if few of the numbers or names are ever contacted, the point is that the teen can see him or herself as *able* to be contacted by a large circle of friends and acquaintances at any time.

Yet for the most part, instant messaging, email, text and multimedia messaging afforded the continuation of conversations that may have begun earlier among peers. Popular cultural artifacts, such as television programs, popular films, or songs, often became the catalyst for discussions of morality that took place both online and off. Like many other teens interviewed, these conversations frequently dealt with issues of sexuality, sexual identity, and sexual orientation, as Steph noted:

> Steph: *I think a lot of times our conversations move so quickly like from what we had for lunch to gay people and lesbian issues and stuff. We are so free because we feel so comfortable and sometimes [in online conversations] we parallel back to something that happened in our daily lives. Like, someone from school said—or we found out that they were bi (sexual) or something so we will bring it up and then we will be "like, oh, did you hear about this thing?" and then everyone will chime in together and might say "well, I don't know if this is right," or "this person was born with this," and someone might say "no, you can't be because I believe that you have to do this." Most of the time it is from our daily lives or sometimes it will be from TV. We were talking about a TV show and they will talk about kissing a bunch of people and then we will just talk about it.*

Steph, who was a regular viewer of several reality television programs on network television and MTV, enjoyed teen-oriented programs such as the WB network's *7th Heaven* and *Gilmore Girls* and syndicated programs such as *Full House*. While she had not looked up any Web sites about these programs online as other teen girls interviewed had, these programs as well as popular music and films often entered her conversations with her friends, particularly when they talked about the purported negative influences of the media on those outside of their friendship circles.

As has been noted in previous research by Baym (2000) and others, email and instant messaging conversations serve to extend conversations among peers that began elsewhere. Often, popular cultural references serve as a bridge for young people to talk through issues of morality, personal experiences, and beliefs, and these references seem to occur as frequently online as off.

Two Patterns of Negotiation

Negotiating with Parents

Consistent with the majority of teens in the U.S., Steph Kline and her parents agreed that she was much more comfortable with computers than her parents were (Lenhart, Rainie, & Lewis, 2001). This generational difference in comfort with technology, as Ribak (2002) has noted, can create problems in intra-family dynamics. While Ribak traced the conflicts that can emerge between boys and their fathers when the younger male's superior technical skills seem to threaten the male authority of the father, Steph's skills do not threaten her parents but rather create new problems of supervision, inconvenience, and expense. As Livingstone (2002) notes, the problem of supervision is a particularly gendered concern for girls using computers in households in which they have more expertise than their parents. For these parents, a teen girl's time with the computer represents psychic time away from family, even as the young person remains in the household. The solution for Steph's family, as well as for many others, echoes a panopticon-like approach to discipline: the implication is that the parent could be looking over her shoulder at any time, and as a result, Steph has learned to police herself.

Steph's parents, like other parents interviewed, noted that the cell phone in particular was originally obtained as a means by which to keep the teen in closer contact with the family. Yet after a period of time, it became more useful for teens as they strove to keep in touch with their friends (see Ling and Yuri, 2002, for discussion of a similar phenomenon in Norway). The technology itself, therefore, becomes a subject for family discord and discussion: what happens when teens block their parents' calls when they are involved in social outings with their friends? What constitutes an "emergency" for either party? And while teen girl monopolization of the household's single phone line is not a new problem, marketers of DSL lines have certainly capitalized on it in recent years. Such decisions involve family negotiations over discipline and surveillance as well as the use of space and time within the home and the allocation of family finances.

Negotiation with Peers

The overriding theme in parent-teen negotiations over new technologies seems to be control of one's environment, specifically as it is related to parental supervision of teen activities. Similarly, the theme that emerges in peer negotiations related to new technologies is the control of one's presentation of oneself. This is perhaps why so many studies of the virtual environment have centered on theories of self-identity (see chapters by Thiel and McMillin in this volume). This study has complemented those by highlighting the negotiations that happen among young people as they seek to manage their offline peer relationships through the online environment. At times, teens in this study noted the ways in which they "hide" behind the computer to afford themselves more security in social situations they perceive as risky, such as raising problems with friends, reaching out to new friends, or making personal revelations that would be more difficult in person. At other times, they employ the technology's written and even its record-keeping aspects to their advantage, revealing others' secrets or quantifying their own popularity relative to their peers. The Internet did not create these practices, of course. Yet creators of technology see in teens a market for products that facilitate these practices. Thus, as the technologies are employed in ways that extend these teen desires for control over parental relations and control in relation to one's self-presentation among peers, they simultaneously meet the social needs of young people and the economic needs of those with technological interests.

Constant Contact: Promises and Fears

Control over one's environment therefore emerges as a central theme in this analysis of teen usage of new media, a control that is negotiated through, and within, the context of constant contact with one's peers. This constant contact, as noted, carries with it both the ability to reduce uncertainties and to engender new social risks. On the one hand, it holds for young women the promise of being able to manage relationships through their written communication skills, which are generally better than those of their male counterparts. Yet on the other hand, the normalizing of such constant contact means the ever-present worry of needing to perform one-self appropriately, and the twin need to be constantly evaluated as acceptable, or simply okay, in the context of one's peers. Trust between peers therefore must be negotiated within each instant message, email, text or multimedia message, or cell phone exchange. This, as Licoppe and Heurtin (2002) point out, is a characteristic of relationships in late modernity:

> Trust must be won by providing tokens of openness, willingness, emotion and commitment. Trust is not therefore a stable category but is constructed as a process, and the bonds it supports become reflexive projects, always open to a revision of the conventions that sustain them. From (Giddens' theory of) the de-institutionalization of personal bonds is borne the necessity repeatedly to display trust towards one another. (p. 107)

Young people use what Ling and Yuri (2002) have termed "hypercoordination" to develop trust in their relationships within the context of expressing themselves. They trust not only that they will receive truthful and self-revelatory communication from their friends and that such truthful communication will be kept confidential, but that they will be made aware of events that are happening among their friends so that they may be included. Being accessible to one's peers thus becomes a central component to being a part of a network built on trust, a network that is cultivated and developed and that simultaneously serves as an expression of one's status relative to one's peers.

Moreover, it is a network that largely exists independently of adult influence, a factor that has long been desirable for this age group. Yet as I have attempted to argue, its online manifestations are not to be understood apart from offline social formations. This is in contrast to Castells' (1996) prediction that an increasingly networked society would see the rise of what he termed the "privatization of sociability," or the rise of online communities as tertiary relationships to primary (familial) and secondary (community) groups. In the phenomenon of the constant contact generation, I argue instead that we are witnessing a rise in wider, geographically located social networks that are characterized by immediate, possibly intimate connections with others and that vary in intensity not by one's association with a particular social subculture or even a particular social group. Instead, we see the emergence of a much more individually organized social network, one that is organized around concerns about how one is able to present oneself in relation to what one perceives as the desired image for different peers. Young people move freely among different communication forms in pursuit of the maintenance of their social network and the sense of self that it bestows on them (see Drotner, 2000).

What are the implications of this drive for constant contact? Employing Goffman's (1959) idea that we each must construct "front stage" versions of ourselves for public consumption, attempting to reserve our inconsistencies and less desirable traits for the "back stage," we might wonder whether the drive toward constant contact between peers further erodes any notions of a "back stage" life separate from the glance of one's peers. With the advent of blogging[12] and the creation of text message "diaries,"[13] it seems that all technologically mediated communication is potentially front stage. This, in fact, constitutes one of the central risks to the new media environment, at least from the perspective of the teen self. Within this environment, teens must learn to negotiate personal privacy issues such as when to elect to not answer instant messages, text messages, and emails, how to respond when your own communication with others goes unanswered, and how to negotiate various transgressions of online friendship norms.

From this study, however, it is possible to surmise that teens have been able to employ new technologies to negotiate privacy issues in relation to their family and have done so quite well. As Livingstone and Bober (2003) have pointed out, "the Internet is a new context between regulation and evasion, surveillance and secrecy, between normative expectation and creative experimentation" (p. 31).

Today's young people experience constant accessibility, separation from adults, and their multitasking abilities as liberating and empowering, a way to manage risk and to direct one's own life course. These technologically mediated communication experiences represent for them an openness to possibilities rather than a limit on their possibilities for privacy, personal reflection, and individual direction, all of which they can also be. Yet we must finally consider the fact that each of these practices is made profitable through the sale of technology to young people. Additionally, it is important to recognize that these purportedly liberating practices look quite different as teens enter an increasingly surveillance-oriented workplace that demands constant accessibility from its workers as well as a well-honed ability to multitask. We must question how these current "fun" practices will be translated into the taken-for-granted understanding of oneself in relation to that workplace, for here we can catch glimpses of how practices experienced as liberating at a certain point in life can engender and maintain a sense of alienation at a later point in life.

This chapter has argued that teen girls, through their uses of new technology, experience themselves as active agents in control of their environments and of their relationships with those who are important to them. Being in constant contact with their peers appeals to them because it presents itself as a way to manage relationships in such a way as to present oneself to others in ways that are highly tailored to the unique situation and audience at hand. Being in constant contact with one's peers has an added benefit as well. When technology is employed as a means by which to keep in touch with peers on the family's home computer, the technology enables young girls to manage parental desires for surveillance and privacy that tend to define their home experience. Due to the limits of space, the chapter can only hint at how these seemingly empowering teen experiences of multitasking and constant accessibility may echo a set of employer-defined expectations that end up erecting limits on leisure and personal privacy at later points in life. Future research could explore this and other related issues longitudinally, so that we might come to a clearer understanding of how being in constant contact with one's peers may continue to serve as an organizing principle for future teens and for the development of technology designed to appeal to them.

Notes

1. Steph Kline and all of the other names mentioned here are pseudonyms.
2. Instant Messaging is a downloadable option provided by many Internet service providers that enables two or more persons who are online in differing locations to communicate with one another using text in real time. When a person opts to instant message another person within the same service provider, a window pops up so that both persons—in addition to others who may be invited to join the conversation—can see the text entries as they are entered by participants. AOL Instant Message (AIM), Yahoo! Messenger, MSN Messenger, and Trillian are popular instant message platforms for teens.
3. Steph was interviewed on two occasions by Monica Emerich, who was a doctoral student and a research assistant on my research project at the University of Colorado at the time of this interview. I am grateful to Monica for her excellent interviewing skills.
4. Short messaging services (SMS) enable cell phone users to send text messages to others without incurring the cost of a cell phone call. Multimedia messages (MMS) are digital photo or video messages sent via cell phone. Both can be used in ways similar to instant messaging, as cell phone users can opt to be notified when they receive a text message, so that a conversation can take place in real time. Because of the limits of the cell phone keypad and screen, these messages are usually shorter than those of instant messages.
5. Annae Gardner was interviewed on two occasions by Diane Alters, who was a doctoral student and a research assistant on my research project at the University of Colorado at the time of this interview.
6. Montana Odell was interviewed on two occasions by Monica Emerich.
7. Seth Rodanski was interviewed on two occasions by Monica Emerich.
8. The interview of this male friendship circle was conducted by Scott Webber, who was a doctoral student and research associate on this project at the University of Colorado during this study.
9. Gabriella Richards was interviewed on two occasions by Monica Emerich.
10. Miranda and Victoria were interviewed in a female friendship circle interview, conducted by the author.
11. Jeremy Friedman was interviewed on two occasions by Denice Walker, a doctoral student and research associate on this research project.
12. Blogging is a form of online journaling that enables individuals to post messages in a public forum. As such, blogging can be a form of alternative journalism and/or an outlet for personal expression. Popular downloadable blogging programs include LiveJournal, Moveable Type, and Blogger.
13. Ling and Yuri (2002) have found that some teen girls keep track of the cell phone-based text message conversations they have had by printing them out and saving them as a "diary" of sorts.

References

Baym, N. (2000). *Tune in, log on: Soaps, fandom, and online community.* Thousand Oaks, CA: Sage.

Buckingham, D. (2000). *After the death of childhood: Growing up in the age of electronic media.* London: Arnold.

Castells, M. (1996). *The rise of the network society.* Cambridge, MA: Blackwell Publishers.

Clark, L. S. (1998). Dating on the "Net": Teens and the rise of "pure" relationships. In S. Jones (Ed.), *Cybersociety 2.0* (pp. 159–183). Thousand Oaks, CA: Sage.

Clark, L.S. (2003, May). Exploring teen friendship networks online: The constant contact generation. Paper presented to the International Communication Association, San Diego, CA.

Corrado, A., & Firestone, C. M. (1996). *Elections in cyberspace: Toward a new era in American politics.* Washington, DC: Aspen Institute.

Dresang, E. T. (1999). More research needed: Informal information-seeking behavior of youth on the Internet. *Journal of the American Society for Information Science, 50*(12), 1123–1124.

Drotner, K. (2000). Difference and diversity: Trends in young Danes' media uses. *Media, Culture, and Society, 22*(2), 149–166.

Facer, K., & Furlong, R. (2001). Beyond the myth of the "Cyberkid": Young people at the margins of the information revolution. *Journal of Youth Studies, 4*(4), 451–469.

Facer, K., Sutherland, R., Furlong, R., & Furlong, J. (2001). What's the point of using computers? The development of young people's computer expertise in the home. *New Media & Society, 3*(2), 199–219.

Giddens, A. (1990). *The consequences of modernity.* Cambridge, UK: Polity Press.

Goffman, E. (1959). *The presentation of self in everyday life.* New York/Garden City: Anchor Press.

Hakken, D. (1999). *Cyborgs @ Cyberspace? An ethnographer looks to the future.* New York: Routledge.

Hakken, D. (2003). *The knowledge landscapes of Cyberspace.* New York: Routledge.

Hirsch, S. (1999). Children's relevance criteria and information seeking on electronic resources. *Journal of the American Society for Information Science, 50*(14), 1265–1283.

Jones, S. (1995). Introduction. In S. Jones (Ed.), *Cybersociety: Computer-mediated communication and community* (pp. 1–35). Thousand Oaks, CA: Sage.

Jones, S. (1998). Introduction. In S. Jones (Ed.), *Cybersociety 2.0: Revisiting CMC and community* (pp. 1–34). Thousand Oaks, CA: Sage.

Jones, S. (2000). The bias of the Web. In A. Herman & T. Swiss (Eds.), *The world wide web and contemporary cultural theory* (pp. 171–182). London: Routledge.

Kasesniemi, E., & Rautianen, P. (2002). Mobile culture of children and teenagers in Finland. In J. Katz & M. Aakhus (Eds.), *Perpetual contact: Mobile communication, private talk, public performance* (pp. 170–192). Cambridge: Cambridge University Press.

Kuhlthau, C. (1991). Inside the search process: Information seeking from the user's perspective. *Journal of the American Society for Information Science, 42*(5), 361–371.

Lenhart, A., Rainie, L., & Lewis, O. (2001). Teenage life online: The rise of the instant-message generation and the Internet's impact on friendships and family relationships. Report from the Pew Internet and American Life Project. Retrieved June 20, 2001, from http://www.pewinternet.org/

Leonardi, P. M. (2003). Problematizing "new media": Culturally based perceptions of cell phones, computers, and the Internet among United States Latinos. *Critical Studies in Media Communication, 20*(2), 160–179.

Licoppe, C., & J. P. Heurtin (2002). France: Preserving the image. In J. Katz & M. Aakhus (Eds.), *Perpetual contact: Mobile communication, private talk, public performance* (pp. 94–109). Cambridge: Cambridge University Press.

Ling, R., & Yuri, B. (2002). Hypercoordination via mobile phones in Norway. In J. Katz & M. Aakhus (Eds.), *Perpetual contact: Mobile communication, private talk, public performance* (pp. 139–169). Cambridge: Cambridge University Press.

Livingstone, S. (2002). *Young people and new media*. London: Sage.

Livingstone, S., & Bober, M. (2003). *UK children go online: Listening to young people's experiences*. London: London School of Economics.

Livingstone, S., & Bovill, M. (1999). *Children, young people, and the changing media environment: Final report*. London: London School of Economics.

McMillin, D. (2005). Teen crossings: Emerging cyberpublics in India. In S. R. Mazzarella (Ed.), *Girl wide web: Girls, the Internet, and the negotiation of identity* (pp. 161–178). New York: Peter Lang.

Norris, P., & Jones, P. (1998). Virtual democracy. *Harvard International Journal of Press/Politics, 3*(2), 1–4.

Ribak, R. (2002). "Like Immigrants": Negotiating power in the face of the home computer. *New Media & Society, 3*(2), 220–239.

Rice, R., & Katz, J. (2004). The Internet and political involvement in 1996 and 2000. In P. Howard & S. Jones (Eds.), *Society online: The Internet in context* (pp. 103–120). New York: Routledge.

Scanlon, M., & Buckingham, D. (2003). Learning online: E-learning and the domestic market in the U.K. In G. Marshall & Y. Katz (Eds.), *Learning in school, home, and community* (pp. 137–148). Boston: Kluwer.

Shade, L. (2002). *Children, young people and new media in the home*. Unpublished project proposal.

Shah, D. V., Kwak, N., & Holbert, R. L. (2001). "Connecting" and "disconnecting" with civic life: Patterns of Internet use and the production of social capital. *Political Communication, 18*, 141–162.

Slevin, J. (2000). *The Internet and society*. Cambridge, UK: Polity Press.

Swidler, A. (1986). Culture in action: Symbols and strategies. *American Sociological Review, 51*(April), 273–86.

Thiel, S. (2005). "IM me": Identity construction and gender negotiation in the world of adolescent girls and Instant Messaging. In S.R. Mazzarella (Ed.), *Girl wide web: Girls, the Internet, and the negotiation of identity* (pp. 179–201). New York: Peter Lang.

Tufte, B. (2003). Girls in the new media landscape. *Nordicom Review, 24*(1), 71–78.

Turkle, S. (1995). *Life on the screen: Identity in the age of the Internet*. New York: Simon & Schuster.

Contributors

Lynn Schofield Clark (Ph.D., University of Colorado) is assistant research professor at the University of Colorado and director of the Teens and the New Media @ Home Project, an ethnographically based research initiative focusing on the use of new media technologies among young people. She teaches courses in critical/cultural studies, including emphases in qualitative audience studies, television, new media, popular culture and popular religion, and the music industry. A former television producer and marketing professional, Clark is author of the critically acclaimed book *From Angels to Aliens: Teenagers, the Media, and the Supernatural* (Oxford UP, 2003), co-author of *Media, Home, and Family* (Routledge, 2003), and co-editor of *Practicing Religion in the Age of the Media* (Columbia UP, 2002). She has published in the *Journal of Communication, Critical Studies in Mass Communication,* and *New Media & Society,* in addition to several other journals and edited volumes.

Lynne Y. Edwards (Ph.D., University of Pennsylvania) is associate professor of Media and Communication Studies at Ursinus College in Collegeville, PA. She is the author of "Slaying in Black and White: Kendra as Tragic Mulatta in Buffy the Vampire Slayer" in *Fighting the Forces: What's at Stake in Buffy the Vampire Slayer* (Rowman & Littlefield, 2002) and "Black Like Me: Value Commitment and Television Viewing Preferences of U.S. Black Teenage Girls" in *Black Marks: Minority Ethnic Audiences and Media* (Ashgate, 2001). She currently is writing a book analyzing print news coverage of juvenile crime during the last decade.

Kimberly S. Gregson (Ph.D., Indiana University) is assistant professor in the Department of Television-Radio at Ithaca College, where she teaches courses in media research. Her recently completed dissertation examined emotional and evaluative responses to extreme media. Prior to joining the faculty at Ithaca College, she taught courses in mass communication at Indiana State University and Indiana University-Bloomington. Her research interests include viewers' responses to extreme media, applying disposition theory to extreme media, and uses and gratifications of niche media such as professional wrestling and anime.

Ashley D. Grisso is a doctoral candidate in Communication at the University of New Mexico where she teaches media, intercultural, and gender and communication courses. Her research interests include the interrelationship of culture, communication, and media, with an emphasis on gender and youth culture. Her dissertation, "Ritualized Use of Media Among Middle School Aged Girls," explores the ways in which media rituals form and inform adolescent female sexuality and spirituality. Informing her current research are her twenty years of "real world" experience, including marketing, consulting, and media-literacy curriculum development and implementation. She can be reached at ashvoice@earthlink.net.

Susan J. Harewood is a doctoral candidate in the Institute of Communication Research at the University of Illinois. Her research interests center upon popular culture, gender and postcolonialism. Her dissertation focuses on the construction and performance of identity in Caribbean popular culture. She has published in *Harvard Educational Review* and has an upcoming article in *Cultural Studies–Critical Methodologies*.

Sharon R. Mazzarella (Ph.D., University of Illinois) is associate professor in the Department of Television and Radio at Ithaca College where she teaches courses in youth culture, media effects, and media studies. Her research focuses on youth culture and mass media. She is founding and lead co-editor of the journal *Popular Communication* (Lawrence Erlbuam Associates) and has published articles in *Popular Music and Society*, the *Journal of Broadcasting & Electronic Media*, and *Communication Research*. She is co-editor of *Growing Up Girls: Popular Culture and the Construction of Identity* (Peter Lang, 1999), has published several book chapters on youth and media, and currently is writing a book on media framing of young people.

Divya C. McMillin (Ph.D., Indiana University) is assistant professor of international communication and cultural studies in the Department of Interdisciplinary Arts and Sciences at the University of Washington, Tacoma. She has published articles in the *Journal of Communication, International Journal of Cultural Studies, Continuum: Journal of Media and Cultural Studies, Indian Journal of Gender Studies,* and the *International Communication Bulletin.* In addition, she has written book chapters, and presented award-winning papers at national and international conferences on issues of globalization, national identity, transnational media networks, and television reception. She currently is working on the book, *International Media Studies* (Blackwell).

Debra Merskin (Ph.D., Syracuse University) is associate professor at the School of Journalism & Communication at the University of Oregon, where she teaches in the communication studies sequence. Her research focuses on media and advertising representations of sexuality, race, and the female body and has been published in such journals as *Howard Journal of Communication, Sex Roles: A Journal of Research, Feminist Media Studies, Journalism & Mass Communication Quarterly,* and *Current Issues in Advertising Research.*

Christine Scodari (Ph.D., Ohio State University) is associate professor of communication and women's studies associate at Florida Atlantic University. She is author of *Serial Monogamy: Soap Opera, Lifespan, and the Gendered Politics of Fantasy* (Hampton, 2004), a project encompassing virtual ethnography of soap opera fans. Her other recent work relating to gender and online fandom has been published in *Popular Communication*, *Communication Studies* and *Women's Studies in Communication*.

Shayla Marie Thiel (Ph.D., University of Iowa) is assistant professor in the Department of Communication at DePaul University in Chicago. She recently completed her doctoral dissertation on adolescent girls' use of instant messaging and identity construction. Her research interests include gender studies and online media. Formerly an online journalist who worked for such publications as washingtonpost.com, *The Chronicle of Higher Education*, and CNN.com, Thiel was the editor of *The Journal of Communication Inquiry* for the 2003–2004 academic year and has published in *Feminist Media Studies* and *The Journal of Electronic Publishing*.

Angharad N. Valdivia (Ph.D., University of Illinois) is research professor at the Institute of Communications Research at the University of Illinois. She is the author of *A Latina in the Land of Hollywood* (University of Arizona Press, 2000) and editor of *Feminism, Multiculturalism and the Media* (Sage, 1995), the Communications and Culture section of the *Routledge International Encyclopedia of Women* (2000), the *Blackwell Companion to Media Studies* (2003), and co-editor of the forthcoming *Geographies of Latinidad: Latina/o Studies into the Twenty-first Century* (Duke UP). Her research and teaching focus on transnational gender issues with a special emphasis on Latinas and Latin America in popular culture.

Susan F. Walsh (Ph.D., University of Oregon) is associate professor in the Department of Communication at Southern Oregon University in Ashland, Oregon. Her teaching and research interests focus on the intersections between gender, communication, and cultural texts, with a particular emphasis on popular culture and the media. She is the co-author of "'Radical' Feminists and 'Bickering' Women: Backlash in U.S. Media Coverage of the United Nations Fourth World Conference on Women" (*Critical Studies in Mass Communication*, 1999, 16:1).

David Weiss (M.A., University of Oregon) is a doctoral candidate in the Department of Communication and Journalism at the University of New Mexico. Prior to his return to academia, David spent nearly two decades in the advertising business and ran his own consultancy specializing in marketing to children. In addition to presenting conference papers on such topics as gender roles in contemporary TV game shows, the place of spirituality in the discourse of cognitive science, and the utopian visions implicit in urban graffiti, he has published a number of literary and journalistic essays in national and regional magazines and newspapers. His dissertation is a critical/ideological analysis of religious discourse and the language(s) of faith in the 2004 presidential election campaigns.

Intersections
in Communications
and Culture

Global Approaches and Transdisciplinary Perspectives

General Editors: Cameron McCarthy & Angharad N. Valdivia

An Institute of Communications Research, University of Illinois Commemorative Series

This series aims to publish a range of new critical scholarship that seeks to engage and transcend the disciplinary isolationism and genre confinement that now characterizes so much of contemporary research in communication studies and related fields. The editors are particularly interested in manuscripts that address the broad intersections, movement, and hybrid trajectories that currently define the encounters between human groups in modern institutions and societies and the way these dynamic intersections are coded and represented in contemporary popular cultural forms and in the organization of knowledge. Works that emphasize methodological nuance, texture and dialogue across traditions and disciplines (communications, feminist studies, area and ethnic studies, arts, humanities, sciences, education, philosophy, etc.) and that engage the dynamics of variation, diversity and discontinuity in the local and international settings are strongly encouraged.

L I S T   O F   T O P I C S

- *Multidisciplinary Media Studies*
- *Cultural Studies*
- *Gender, Race, & Class*
- *Postcolonialism*
- *Globalization*
- *Diaspora Studies*
- *Border Studies*
- *Popular Culture*
- *Art & Representation*
- *Body Politics*
- *Governing Practices*

- *Histories of the Present*
- *Health (Policy) Studies*
- *Space and Identity*
- *(Im)migration*
- *Global Ethnographies*
- *Public Intellectuals*
- *World Music*
- *Virtual Identity Studies*
- *Queer Theory*
- *Critical Multiculturalism*

Manuscripts should be sent to:

Cameron McCarthy OR **Angharad N. Valdivia**
Institute of Communications Research
University of Illinois at Urbana-Champaign
222B Armory Bldg., 555 E. Armory Avenue
Champaign, IL 61820

To order other books in this series, please contact our Customer Service Department:
    *(800) 770-*LANG *(within the U.S.)*
    (212) 647-7706 (outside the U.S.)
    (212) 647-7707 FAX

Or browse online by series:
    w w w . p e t e r l a n g . c o m